Estimation of Stream Temperature in Support of Fish Production Modeling under Future Climates in the Klamath River Basin

By Lorraine E. Flint and Alan L. Flint

Prepared in cooperation with the U.S. Bureau of Reclamation and U.S. Fish and Wildlife Service

Scientific Investigations Report 2011–5171

U.S. Department of the Interior
U.S. Geological Survey

U.S. Department of the Interior
KEN SALAZAR, Secretary

U.S. Geological Survey
Marcia K. McNutt, Director

U.S. Geological Survey, Reston, Virginia: 2012

For more information on the USGS—the Federal source for science about the Earth, its natural and living resources, natural hazards, and the environment, visit http://www.usgs.gov or call 1–888–ASK–USGS.

For an overview of USGS information products, including maps, imagery, and publications, visit http://www.usgs.gov/pubprod

To order this and other USGS information products, visit http://store.usgs.gov

Contents

Figures

Tables

Conversion Factors and Datum

Multiply	By	To obtain
Length		
inch (in)	2.54	centimeter (cm)
foot (ft)	0.3048	meter (m)
mile (mi)	1.609	kilometer (km)
Area		
acre	4,047	square meter (m^2)
acre	0.4047	hectare (ha)
square mile (mi^2)	259.0	hectare (ha)
Volume		
cubic foot (ft^3)	0.028317	cubic meter
acre-foot (acre-ft)	1,233	cubic meter (m^3)
Flow rate		
cubic foot per second (ft^3/s)	0.02832	cubic meter per second
million gallons per day (Mgal/d)	0.04381	cubic meter per second
inch per hour (in/h)	0.0254	meter per hour (m/h)
inch per year (in/yr)	2.54	centimeter per year (cm/yr)
mile per hour (mi/h)	1.609	kilometer per hour (km/h)
Mass		
ounce, avoirdupois (oz)	28.35	gram (g)
pound, avoirdupois (lb)	0.4536	kilogram (kg)
pound per acre (lb/acre)	1.121	kilogram per hectare

Temperature in degrees Celsius (°C) may be converted to degrees Fahrenheit (°F) as follows:

$$°F = (1.8 \times °C) + 32$$

Horizontal coordinate information is referenced to the North American Datum of 1927 (NAD 27).

Estimation of Stream Temperature in Support of Fish Production Modeling under Future Climates in the Klamath River Basin

By Lorraine E. Flint and Alan L. Flint

Abstract

Stream temperature estimates under future climatic conditions were needed in support of fish production modeling for evaluation of effects of dam removal in the Klamath River Basin. To allow for the persistence of the Klamath River salmon fishery, an upcoming Secretarial Determination in 2012 will review potential changes in water quality and stream temperature to assess alternative scenarios, including dam removal. Daily stream temperature models were developed by using a regression model approach with simulated net solar radiation, vapor density deficit calculated on the basis of air temperature, and mean daily air temperature. Models were calibrated for 6 streams in the Lower, and 18 streams in the Upper, Klamath Basin by using measured stream temperatures for 1999–2008. The standard error of the y-estimate for the estimation of stream temperature for the 24 streams ranged from 0.36 to 1.64 degrees Celsius (°C), with an average error of 1.12°C for all streams. The regression models were then used with projected air temperatures to estimate future stream temperatures for 2010–99. Although the mean change from the baseline historical period of 1950–99 to the projected future period of 2070–99 is only 1.2°C, it ranges from 3.4°C for the Shasta River to no change for Fall Creek and Trout Creek. Variability is also evident in the future with a mean change in temperature for all streams from the baseline period to the projected period of 2070–99 of only 1°C, while the range in stream temperature change is from 0 to 2.1°C. The baseline period, 1950–99, to which the air temperature projections were corrected, established the starting point for the projected changes in air temperature. The average measured daily air temperature for the calibration period 1999–2008, however, was found to be as much as 2.3°C higher than baseline for some rivers, indicating that warming conditions have already occurred in many areas of the Klamath River Basin, and that the stream temperature projections for the 21st century could be underestimating the actual change.

Introduction

Salmonids in the Klamath River Basin are currently experiencing multiple stresses from both anthropogenic and natural sources, and have much at stake in the potential outcomes of the pending Secretarial Determination (SD) in 2012 to either remove or retain four hydropower dams in the upper basin below Upper Klamath Lake (*fig. 1*). Salmonids are living at an ecological edge for thermal conditions (Bartholow, 2005), and are already at risk within an environment that is affected by local water management, and that is experiencing underlying temperature and hydrologic changes as a result of global climate change. Water temperature in the Klamath Basin has already increased by about 0.5°C per decade over the period of 1962–2003 (Bartholow and others, 2005). Additional environmental stressors to the Klamath Basin caused by climate change include extended summer low flows and earlier seasonal rises in springtime stream temperatures. These changes in stream temperature are not conducive for survival of cold water species, such as salmonids, that exploit the available thermal extremes for life cycle needs that are present in the spring when water temperature is warming, and in the fall when water temperature is cooling (Shuter and Meisner, 1992).

As part of the upcoming SD in 2012, the effects of dam removal will have to be evaluated in the context of various climate change scenarios. Predictions of future temperature increases vary depending on the General Circulation Model (GCM) or regional model used. The models generally agree that air temperature will increase (Barnett and others, 2004; Payne and others, 2004; Maurer and Duffy, 2005; Vicuna and Dracup, 2007; Brekke and others, 2009), with a potential range of increase in air temperature of 1.5 to 4.5°C by 2100 for California (Cayan and others, 2008).

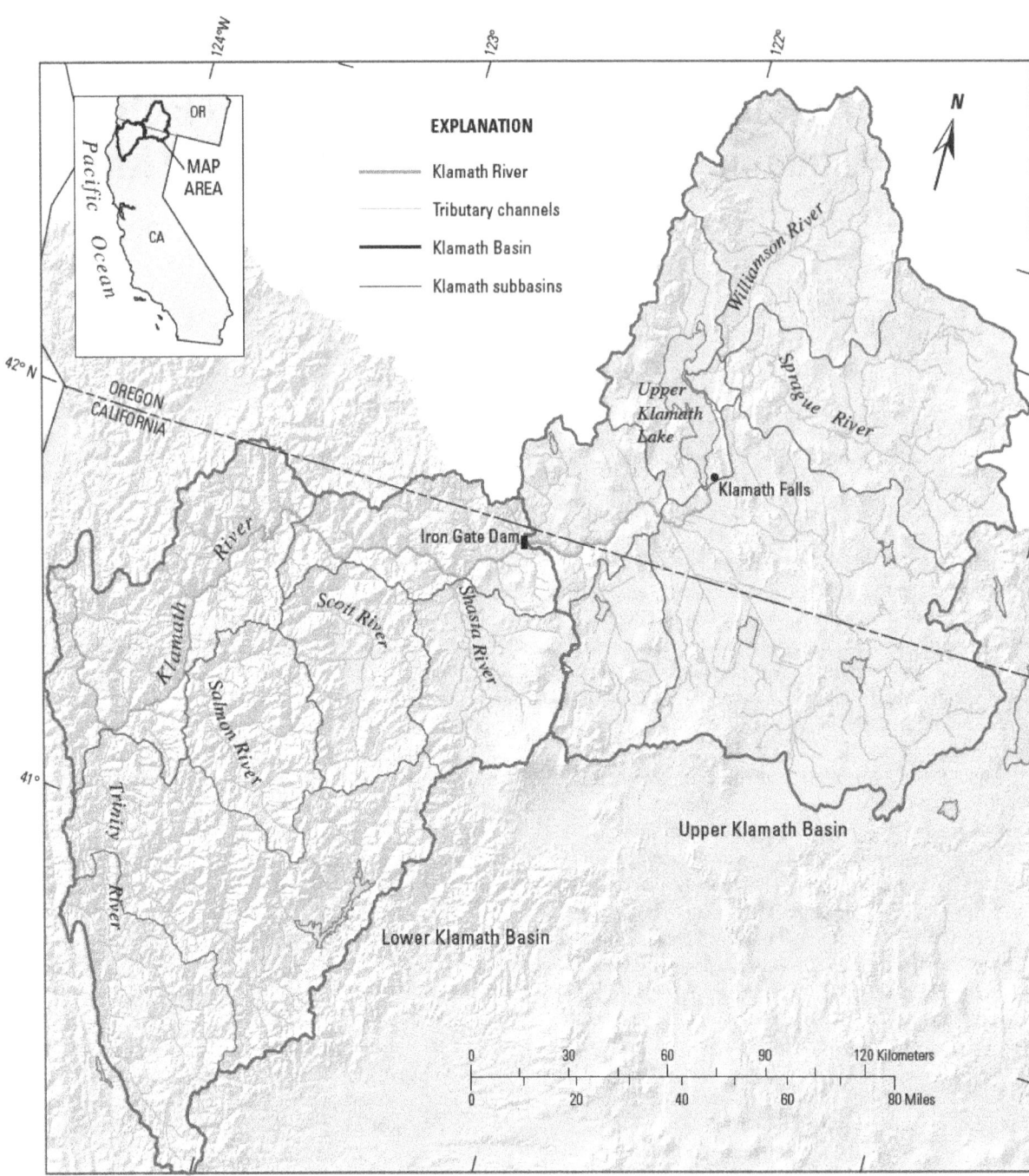

Figure 1.   Klamath River Basin and major tributaries.

Patterns of precipitation and runoff are also likely to change over the next century, although the direction of change for precipitation is much less certain. For northern California, little change is projected during the 21st century, but there is a tendency for increases in the numbers and magnitudes of large precipitation events (Cayan and others, 2008). Although the average annual runoff in many river basins could stay relatively stable, the timing of runoff and the form of precipitation is likely to change as a result of warmer winters. Snow-melt driven basins will receive peak runoff earlier in the year (Barnett and others, 2004). Baseflow could be lower in summer, and could become more dependent on reservoir storage in regulated rivers in order to meet instream flows mandated by biological opinions (Payne and others, 2004) by the U.S. Fish and Wildlife Service (FWS) or the National Marine Fisheries Service (NMFS) program of the National Oceanic and Atmospheric Administration (NOAA).

Resource managers are seeking both short- and long-term mitigation and management options that can allow the Klamath River salmon fishery to persist and improve in the future. Thermal refugia and restored habitat would provide more physical space and access to more tributaries to the Klamath River for stream spawning salmonids (National Marine Fisheries Service, 2006). Dam removal on the Klamath could provide longer spring/fall periods when temperature is in the preferred range for the rearing/spawning life stages of these fish, as well as providing access to additional suitable rearing/spawning habitat. Under the SD process, the best scientific evidence is required to estimate potential effects of dam removal or retention on Klamath Basin fisheries. Simulated future effects of dam removal and ongoing changes in climate are necessary for assessing alternative scenarios. Therefore, modeling tools will play an important role in evaluating dam removal as a feasible option for meeting desired resource management objectives. Existing decision support system (DSS) models for the SD include SIAM/SALMOD (Bartholow and others, 2005), but it is also important to begin laying the foundation for migrating the concepts of SIAM/SALMOD to a more advanced modeling framework for the future. These modeling tools require detailed input data for current climatic conditions for calibration purposes, as well as for potential future projected climatic conditions. Potential changes in air temperature at reasonable spatial scales can greatly influence stream temperature and provide a dominant driver for water-quality simulations for decision support systems for addressing potential effects of dam removal in the Klamath Basin.

Purpose and Scope

The purpose of this report is to document work being done to derive the potential future stream temperatures necessary as input to water-quality models for the SIAM/SALMOD decision support system using future climate projections. This work was performed in support of research providing scientific input to the SD in 2012. Stream temperature models were developed, calibrated to measured stream temperatures, and a extrapolated using future projections of air temperature for multiple 21st century scenarios.

Description of the Study Area

The Klamath River Basin spans the Oregon-California border and ranges from high elevation, relatively flat volcanic deposits in the upper basin, to steep, dissected river channels in the lower basin. Upper Klamath Lake, the largest natural lake in Oregon, is fed primarily by the Sprague, Williamson, and Wood Rivers as well as numerous springs that flow directly into the lake. Water flowing out of Upper Klamath Lake becomes the Klamath River, which flows 423 river kilometers (km) to the Pacific Ocean, cutting through the Cascade and coastal mountain ranges. The basin drains approximately 21,000 square km and encompasses parts of three Oregon and five California counties.

Climate patterns are very variable throughout the Klamath Basin, ranging from varying degrees of marine influences in the coastal region, with moderated temperatures and higher precipitation, to warmer, drier summers and winter snowpack in the upper basin. Generally, temperatures are warmest in July and coolest in January. December and January are the wettest months, and July is the driest. In Klamath Falls, in the Upper Klamath Basin, annual average high temperature is 16°C, and the average low is 2°C. Average January temperatures range between −6°C and 3°C, while July temperatures range between 11°C and 30°C. Klamath Falls receives about 340 millimeters (mm) of rain each year. January and December are the wettest months (50 mm per month), and July is the driest (10 mm). Annual average high temperature is also 16°C, but the average low is 7°C in Klamath, California, near the mouth of the Klamath River. Average minimum and maximum January temperatures are 3°C and 12°C, while July averages are 11°C and 19°C. The months of December and January each receive about 360 mm of rain in Klamath, and the yearly total is 2,030 mm; July is very dry, in contrast, getting only about 10 mm. Springtime snowmelt from the tributaries contributes to high flows from April to June, providing cold water and thermal refugia. By late summer and early fall, flows are typically low and thermal refugia have diminished significantly in most locations in the main stem Klamath River.

Methods

The approach used to estimate stream temperatures for the Klamath River Basin is based on methods developed by Flint and Flint (2008) that developed multiple regression models for streams by using measured stream-temperature data, measured air temperature and relative humidity data, and simulated net solar radiation.

Development of Future Climate Scenarios

Future climate projections were developed by U.S. Bureau of Reclamation (Raff, 2009) by using climate projections housed in a downscaled climate projection archive (*http://gdo-dcp.ucllnl.org/downscaled_cmip3_projections/*). On the basis of criteria described by Raff (2009), 112 climate projections from general circulation models (GCMs) were downscaled to 1/8 degree spatial resolution and evaluated for 1950 through 2099. These represent greenhouse gas emissions paths, A1B, A2, and B1 (International Panel on Climate Change, 2000) for 16 different GCMs and different initial conditions for different model simulations. The GCMs embody the efforts of various climate modeling groups coordinating through the World Climate Research Programme Working Group on Coupled Modeling through the Coupled Model Intercomparison Project Phase 3 (CMIP3) effort (see Meehl and others, 2007). The downscaling from the original GCM scale of 2 degrees spatial resolution, which represents a spatial scale that is too coarse for most impact studies, was done following a methodology by Wood and others (2004) and was applied to these projections to provide information at a 1/8 degree resolution that can be used to study potential climate change effects.

Of the three emissions scenarios within the downscaled archive, 75 projections were extracted representing all of the projections following the A1B and A2 emissions paths. The A1B and A2 scenarios use higher greenhouse gas emissions than the B1 emissions scenario. The B1 emissions scenario was not included because global emissions are already known to exceed all scenarios described in the Intergovernmental Panel on Climate Change (IPCC) Special Report on Emission Scenarios (SRES; IPCC, 2000) at the present time, and, therefore, the B1 projections were considered less likely future projections than the A1B and A2 projections.

To correspond with existing tools and methods for evaluating water quality in the region, projections were divided into three regions: the Upper Basin, encompassed by the region upstream of Iron Gate Dam (*fig. 2*); the Lower Basin, encompassing all basins downstream of Iron Gate Dam; and the Coastal Basin, encompassing the furthest downstream reaches of the Klamath and Trinity Rivers. Temporally, and strictly for development of the bias-corrected projections, the projections were divided into two equal length periods—the baseline period defined as 1950–99, and a 'lookahead' period defined as 2020–69. The lookahead period was chosen on the basis of the analysis period defined for the Klamath Dam Removal Study (Reclamation, 2011). A 50-year baseline and a lookahead period were used to encompass the full time period of analysis that would lead to a single set of projections to facilitate the evaluation. The projections for each of the Upper, Lower, and Coastal regions of the Klamath Basin were averaged both spatially and over the temporal period. The result of this averaging is a single value of air temperature for each projection for each of the three regions within the baseline period and the lookahead period.

The metrics of climate change that were evaluated were changes in air temperature, described as a net change from baseline to future. Selection of climate-change scenarios were defined by the distribution of the projected net change in air temperature and were distributed within Weibull plots with 25th and 75th quantiles defining four of the scenarios, and the 5th scenario defined as the 50th quantile. Additional details describing the selection of the five scenarios are described in Raff (2009). Ten realizations were done by using the SAC-SMASnow17 hydrology model, which was used to translate the regional climate change scenarios into runoff for the water quality modeling (U.S. Department of Interior, 2009), to incorporate some of the randomness of the temporal scaling technique for each scenario. One realization was chosen that produced the median change in mean-annual unregulated runoff. Thus, five scenarios were made available from Reclamation for 1950–2099 (*table 1*), provided as daily average values of air temperature for each of the three regions.

Data Collection and Development of Model Inputs

Data used in this study for model development were from water years 1999–2008, and included daily maximum and minimum air temperature, and daily relative humidity, all of which were used to calculate vapor density deficit. Measured stream temperatures and simulated net radiation also were used to calibrate the stream temperature regression models.

Meteorology Data and Processing

To best represent the spatial structure of large-scale synoptic meteorological processes, a large number of data stations were analyzed (*fig. 2*; *table 2*). Maximum and minimum daily air temperatures were obtained for 212 stations, and relative humidity for 68 of those stations, from the National Climatic Data Center (NCDC; National Climatic Data Center, 2009) and Remote Automated Weather Stations (RAWS) in and around the Klamath Basin (Western Regional Climate Center; *http://www.raws.dri.edu*; accessed 2010).

Table 1.   Downscaled global climate models used for projecting future stream temperatures (from Raff, 2009) in the Klamath River Basin

Identification	Model
Run 6	cccma_cgcm3_1.4.sresa1b
Run 11	gfdl_cm2_0.1.sresa2
Run 24	miub_echo_g.3.sresa1b
Run 37	mri_cgcm2_3_2a.3.sresa1b
Run 45	ncar_pcm1.1.sresa2

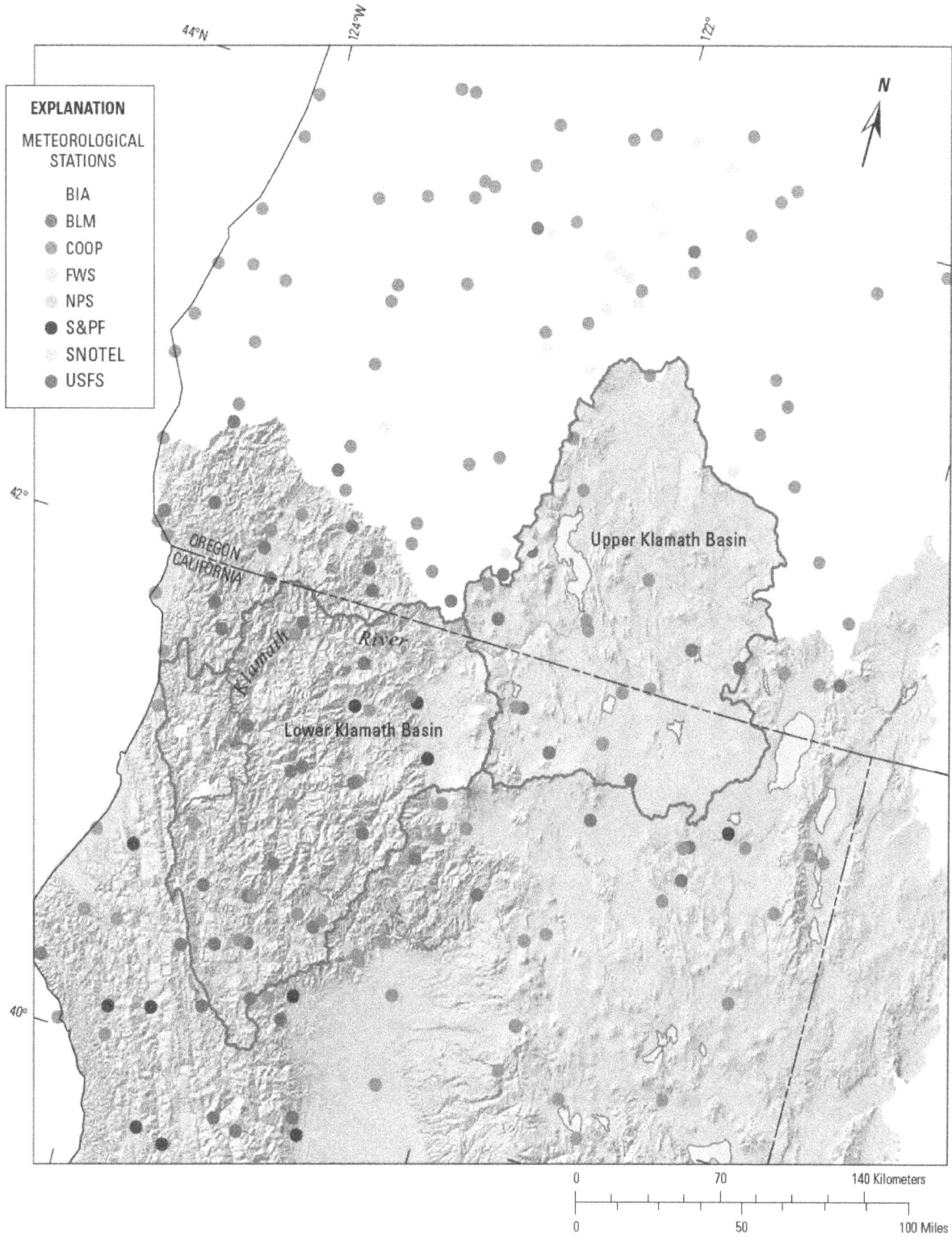

Figure 2. Meteorological stations in and around the Klamath River Basin. (BIA, Bureau of Indian Affairs; BLM, Bureau of Land Management; COOP, National Climatic Data Center cooperative stations; FWS, Fish and Wildlife Service; NPS, National Park Service; RAWS, Remote Automated Weather Stations; SNOTEL, National Resource Conservation Service snow stations; USFS, U. S. Forest Service)

Table 2.   Locations of meterological stations used for air temperature and relative humidity measurements in and around the Klamath River Basin.

Station identification	Name	Latitude, decimal degrees	Longitude, decimal degrees	Elevation, meters
100	Rodeo Valley, California	39.6681	−123.3211	740.1
101	Laytonville	39.7022	−123.4850	554.7
102	Mendocino Pass, California	39.8075	−122.9450	1,652.0
103	Eel River	39.8333	−123.0833	457.2
104	Thomes Creek, California	39.8644	−122.6097	317.0
105	Eagle Peak	39.9278	−122.6569	1,131.7
106	Richardson Grove State Park	40.0261	−123.7931	153.3
107	Shelter Cove Aviation	40.0331	−124.0728	75.0
108	Eel River Camp	40.1383	−123.8236	135.9
109	Red Bluff Municipal Airport	40.1519	−122.2536	107.6
110	Canyon Dam	40.1706	−121.0886	1,389.9
111	Alder Point	40.1867	−123.5903	281.3
112	Ruth, California	40.2506	−123.3158	832.7
113	Cooskie Mountain	40.2569	−124.2661	899.2
114	Patty Mocus, California	40.2950	−122.8667	1,066.8
115	Chester	40.3033	−121.2422	1,380.7
116	Yolla Bolla, California	40.3383	−123.0650	2,059.5
117	Mineral	40.3458	−121.6092	1,485.9
118	Harrison Gulch Reservoir	40.3636	−122.9650	838.2
119	Arbuckle Basin	40.3983	−122.8333	579.1
120	Susanville	40.4167	−120.6631	1,275.3
121	Mad River, California	40.4633	−123.5239	845.8
122	Scotia	40.4831	−124.1036	41.5
123	Grizzly Creek State Park	40.4864	−123.9089	125.9
124	Friend Mountain	40.5050	−123.3417	1,219.2
125	Redding Municipal Airport	40.5175	−122.2986	151.5
126	Manzanita Lake, California	40.5400	−121.5803	1,725.2
127	Manzanita	40.5419	−121.5764	1,752.6
128	Hayfork	40.5500	−123.1650	708.1
129	Hayfork 2	40.5525	−123.2122	701.0
130	Whiskeytown Reservoir	40.6117	−122.5281	394.7
131	Whiskeytown	40.6333	−122.5500	331.9
132	Oak Bottom, California	40.6506	−122.6056	404.2
133	Lowden	40.6894	−122.8314	951.0
134	Shasta Dam	40.7142	−122.4161	327.7
135	Underwood, California	40.7219	−123.4953	792.5
136	Weaverville	40.7222	−122.9331	599.8
137	Trinity River	40.7264	−122.7947	567.2
138	Big Bar	40.7333	−123.2333	457.2
139	1 West northwest Big Bar 4 East	40.7403	−123.2081	381.9
140	Maple Creek, California	40.7964	−123.9367	512.1
141	Regional Airport Eureka	40.8097	−124.1603	6.1
142	Termo 1 East	40.8667	−120.4333	1,615.4

Table 2. Locations of meterological stations used for air temperature and relative humidity measurements in and around the Klamath River Basin.—Continued

Station identification	Name	Latitude, decimal degrees	Longitude, decimal degrees	Elevation, meters
143	Burney	40.8803	−121.6547	974.8
144	Backbone	40.8892	−123.1422	1,432.6
145	Lake Hat Creek	40.9317	−121.5433	919.0
146	Willow Creek 1 Northwest	40.9467	−123.6367	141.4
147	Oak Mountain, California	41.0064	−121.9833	518.2
148	Hoopa	41.0478	−123.6714	114.3
149	Sims, California	41.0750	−122.3733	731.5
150	Big Hill	41.0975	−123.6358	1,088.1
151	Scorpion, California	41.1117	−122.6967	1,341.1
152	School House, California	41.1383	−123.9056	804.7
153	Cecilville	41.1417	−123.1392	704.1
154	Dunsmuir Treatment Plant	41.1833	−122.2736	661.4
155	Adin Reservation	41.1936	−120.9447	1,278.6
156	Adin Mountain	41.2333	−120.7833	1,886.7
157	McCloud	41.2514	−122.1383	999.7
158	Jess Valley	41.2683	−120.2947	1,645.9
159	Blue Ridge	41.2694	−123.1875	1,815.4
160	Ash Creek	41.2769	−121.9794	975.4
161	Yurok, California	41.2897	−123.8575	150.9
162	Rush Creek, California	41.2944	−120.8639	1,463.0
163	Callahan 2	41.3000	−122.8244	1,192.1
164	Sawyers Bar, California	41.3003	−123.1322	668.1
165	Orleans	41.3089	−123.5322	122.8
166	Callahan	41.3111	−122.8044	970.8
167	Mount Shasta	41.3206	−122.3081	1,094.2
168	Orick Prairie Creek Park	41.3619	−124.0192	48.8
169	Somes Bar, California	41.3900	−123.4958	280.4
170	Canby 3 Southwest	41.4219	−120.9017	1,313.7
171	Round Mountain	41.4272	−121.4639	1,602.6
172	Canby	41.4342	−120.8678	1,314.3
173	Weed Airport, California	41.4789	−122.4539	893.1
174	Alturas	41.4931	−120.5528	1,341.1
175	Fort Bidwell	41.5200	−120.0900	1,442.3
176	Devils Garden	41.5300	−120.6714	1,530.7
177	Cedarville	41.5336	−120.1736	1,423.4
178	Cedar Pass	41.5667	−120.3000	2,142.7
179	Klamath	41.5786	−124.0747	8.5
180	Quartz Hill, California	41.5992	−122.9336	1,287.8
181	Fort Jones Reservoir	41.6000	−122.8478	830.6
182	Timber Mountain, California	41.6294	−121.2981	1,511.8
183	Van Bremmer, California	41.6431	−121.7939	1,502.1
184	Brazie Ranch	41.6853	−122.5942	914.4
185	Yreka	41.7036	−122.6408	800.1

Table 2.   Locations of meterological stations used for air temperature and relative humidity measurements in and around the Klamath River Basin.—Continued

Station identification	Name	Latitude, decimal degrees	Longitude, decimal degrees	Elevation, meters
186	Ship Mountain, California	41.7358	–123.7917	1,615.4
187	Lava Beds National Monument	41.7400	–121.5067	1,453.9
188	Indian Well	41.7417	–121.5383	1,453.9
189	Collins Baldy	41.7750	–122.9503	1,674.3
190	Mount Hebron Range Station	41.7836	–122.0447	1,295.4
191	Juanita Lake	41.7861	–122.0056	1,645.9
192	Crescent City 3 North-northwest	41.7958	–124.2147	13.1
193	West Happy Camp Reservoir	41.8042	–123.3758	341.4
194	Camp Six	41.8308	–123.8764	1,151.5
195	Slater Butte, California	41.8586	–123.3525	1,423.4
196	Crowder Flat	41.8833	–120.7500	1,575.8
197	Tulelake	41.9600	–121.4744	1,229.9
198	Dismal Swamp	41.9833	–120.1667	2,243.3
199	Crazy Peak	41.9919	–123.6036	1,210.1
200	Lower Klamath	41.9992	–121.7003	1,249.1
201	Malin 5 East	42.0078	–121.3186	1,410.3
202	Brookings	42.0300	–124.2453	15.2
203	Big Red Mountain	42.0500	–122.8500	1,844.0
204	Squaw Peak, Oregon	42.0667	–123.0167	1,513.0
205	Bigelow Camp	42.0667	–123.3333	1,563.6
206	2 Southeast Brookings	42.0769	–124.3178	53.3
207	Illinois Valley Airport, Oregon	42.1039	–123.6853	423.4
208	Parker Mountain, Oregon	42.1058	–122.2781	1,609.3
209	Strawberry	42.1167	–120.8333	1,758.7
210	Buckhorn Springs, Oregon	42.1197	–122.5633	847.3
211	Red Mound, Oregon	42.1233	–124.3006	534.3
212	Star, Oregon	42.1500	–123.0667	510.8
213	Klamath Falls	42.1644	–121.7547	1,247.2
214	Cave Junction	42.1769	–123.6753	390.1
215	Strawberry, Oregon	42.1894	–120.8464	1,703.8
216	Gerber Reservoir	42.2000	–121.1333	1,490.5
217	Klamath Falls 2 South-southwest	42.2008	–121.7814	1,249.1
218	Gerber, Oregon	42.2056	–121.1389	1,499.6
219	Ashland	42.2128	–122.7144	532.2
220	Lakeview	42.2139	–120.3636	1,456.3
221	Quail Prairie Lookout, Oregon	42.2167	–124.0333	924.5
222	Bly 4 Southeast	42.2208	–120.5792	1,389.9
223	Ruch	42.2231	–123.0472	472.4
224	Howard Prairie	42.2292	–122.3814	1,392.0
225	Summit, Oregon	42.2322	–120.2456	1,873.6
226	Selma 4 East	42.2753	–123.5281	445.0
227	Dead Indian, Oregon	42.2833	–122.3167	1,493.5
228	Provolt Seed Orchard, Oregon	42.2897	–123.2303	359.7

Table 2.   Locations of meterological stations used for air temperature and relative humidity measurements in and around the Klamath River Basin.—Continued

Station identification	Name	Latitude, decimal degrees	Longitude, decimal degrees	Elevation, meters
229	Medford Experimental Station	42.2961	−122.8700	444.1
230	Quartz Mountain	42.3167	−120.8167	1,743.5
231	Fish Lake	42.3667	−122.3333	1,420.4
232	Medford Rogue	42.3811	−122.8722	395.3
233	Billie Creek Divide	42.4000	−122.2500	1,609.3
234	Gold Beach	42.4036	−124.4242	15.2
235	Seldom Creek, Oregon	42.4075	−122.1914	1,485.9
236	Grants Pass	42.4244	−123.3236	283.5
237	Sprague River 2 Southeast	42.4306	−121.4892	1,366.4
238	Fourmile Lake	42.4333	−122.2167	1,819.7
239	Valley Falls	42.4844	−120.2822	1,318.3
240	Merlin Land Fill, Oregon	42.4947	−123.3972	378.0
241	Cold Springs Camp	42.5167	−122.1667	1,810.5
242	Agness, Oregon	42.5522	−124.0578	75.3
243	Sexton Summit	42.6003	−123.3642	1,168.0
244	Illahe	42.6286	−124.0575	106.1
245	Crazyman Flat	42.6333	−120.9333	1,883.7
246	Lost Creek	42.6722	−122.6750	481.6
247	Sevenmile Marsh	42.6833	−122.1333	1,737.4
248	Taylor Butte	42.6833	−121.4167	1,533.1
249	Summer Rim	42.6833	−120.8000	2,158.0
250	Paisley	42.6922	−120.5403	1,328.9
251	Chiloquin	42.7036	−121.9953	1,274.1
252	King Mountain	42.7167	−123.2000	1,322.8
253	Prospect	42.7342	−122.5164	756.5
254	Port Orford 2	42.7519	−124.5011	12.8
255	Sun Pass	42.7833	−121.9667	1,645.9
256	Annie Springs	42.8667	−122.1500	1,831.8
257	Powers	42.8886	−124.0689	70.1
258	Crater Lake	42.8967	−122.1328	1,973.6
259	Langlois	42.9242	−124.4533	27.4
260	Silver Creek	42.9500	−121.1667	1,749.6
261	Riddle	42.9506	−123.3572	207.3
262	Summer Lake	42.9592	−120.7897	1,277.7
263	Silver Lake	43.1244	−121.0619	1,335.6
264	Bandon 2	43.1497	−124.4019	6.1
265	Dora 2 West	43.1639	−123.9956	29.0
266	Diamond Lake	43.1833	−122.1333	1,609.3
267	Coquille City	43.1872	−124.2025	7.0
268	Roseburg KQEN	43.2131	−123.3658	129.5
269	Toketee Airstrip	43.2167	−122.4167	987.6
270	Chemult Alternate	43.2167	−121.8000	1,478.3
271	Chemult	43.2292	−121.7894	1,450.8

Table 2.   Locations of meterological stations used for air temperature and relative humidity measurements in and around the Klamath River Basin.—Continued

Station identification	Name	Latitude, decimal degrees	Longitude, decimal degrees	Elevation, meters
272	The Poplars	43.2644	−120.9447	1,313.7
273	Toketee Falls	43.2750	−122.4497	627.9
274	Winchester	43.2828	−123.3536	140.2
275	Fort Rock	43.3572	−121.0517	1,318.6
276	Lemolo Lake	43.3597	−122.2208	1,242.7
277	Idleyld Park 4 Northeast	43.3708	−122.9653	329.2
278	North Bend	43.4133	−124.2436	1.8
279	Summit Lake	43.4333	−122.1333	1,709.9
280	New Crescent Lake	43.5000	−121.9667	1,496.6
281	Odell Lake–East	43.5492	−121.9639	1,463.0
282	Cascade Summit	43.5833	−122.0500	1,554.5
283	Elkton 3	43.5992	−123.5992	36.6
284	Salt Creek Falls	43.6000	−122.1167	1,286.3
285	Railroad Overpass	43.6500	−122.2000	816.9
286	Drain	43.6656	−123.3275	89.0
287	Holland Meadows	43.6667	−122.5667	1,502.7
288	Sugarloaf, Oregon	43.6728	−122.6564	1,082.0
289	Wickiup Dam	43.6825	−121.6875	1,328.3
290	Cottage Grove Dam	43.7178	−123.0578	253.3
291	Oakridge Fish Hatchery	43.7428	−122.4433	388.6
292	Ranger Station Gardiner	43.7464	−124.1217	9.1
293	Round Mountain, Oregon	43.7639	−121.7167	1,798.3
294	Dorena	43.7822	−122.9631	249.9
295	Cottage Grove 1 North-northeast	43.7917	−123.0275	181.4
296	Irish Taylor	43.8000	−121.9333	1,688.6
297	Brothers	43.8094	−120.6000	1,414.3
298	Sunriver	43.8933	−121.4117	1,274.1
299	Roaring River	43.9000	−122.0167	1,508.8
300	Lookout Point Dam	43.9144	−122.7600	217.0
301	Honeyman State Park	43.9281	−124.1069	35.1
302	Barnes Station	43.9456	−120.2169	1,210.1
303	Bend	44.0569	−121.2850	1,115.6
304	8 North Leaburg 1 Southwest	44.1014	−122.6886	205.7
305	Bend 7 Northeast	44.1183	−121.2103	1,023.5
306	Fern Ridge Dam	44.1236	−123.3064	147.8
307	Eugene Mahlon Sweet Airport	44.1278	−123.2206	107.6
308	Cougar Dam	44.1308	−122.2419	384.0
309	Three Creeks Meadow	44.1333	−121.6333	1,734.3
310	Mckenzie Bridge Reservoir	44.1781	−122.1156	450.5
311	Mckenzie	44.2000	−121.8667	1,453.9

The values of maximum and minimum air temperature and relative humidity were spatially distributed to all grid cells (270 by 270 meters) for the Klamath River Basin model domain for each day by using an equation developed by Nalder and Wein (1998) and modified by Flint and Flint (2008). The equation uses multiple regressions to combine a spatial and elevation gradient with an application of inverse-distance squared weighting of daily point data to interpolate temperature or relative humidity to each grid cell (see Flint and Flint, 2008, fig. 3).

Solar Radiation and Vapor Density Deficit

It was determined by Flint and Flint (2008) that net radiation and vapor density deficit were highly correlated to stream temperature, along with air temperature. Net radiation was simulated following methods described in Flint and Flint (2008). Vapor density deficit is the ratio of vapor density at saturation for the specified air temperature and the current density at the same air temperature. This was used because vapor density deficit, rather than relative humidity, is a major driving force for evaporation or evaporative cooling (Campbell, 1979; Mohseni and Stefan, 1999). Vapor density deficit was calculated from daily mean air temperature and relative humidity by using formulae from Campbell (1979). Vapor density deficit was calculated from relative humidity (RH) and mean air temperature (Tmean, in °C) following Campbell and Norman (1998):

$$esTmean = 0.611 * EXP((17.502 * Tmean) / \qquad \\ (Tmean + 240.97)) \qquad (1)$$

$$VPD = esTmean * (1 \ (RH / 100)) \qquad (2)$$

$$VDD = VPD / (0.000466 * (Tmean + 273.15)) \qquad (3)$$

where

   $esTmean$ is the mean air temperature at saturation,

   VPD is the vapor pressure deficit in kilopascals, and

   VDD is the vapor density deficit in millibars.

Stream Temperature Data and Model Development

Stream temperature data were collected from several sources for 18 tributary streams in the Upper Basin and 6 tributary streams for the Lower Basin. Locations and sources are shown in *table 3*. In order to delineate the upstream area for contributing streamflows, all grid cells upstream of each measurement location were identified by using the USGS National Hydrography Dataset and ArcGIS (ESRI; *www.esri.com/arcgis*).

To develop the representative parameters for the stream temperature analysis, the data were extracted only for grid cells that intersected stream channels because it was assumed that the higher elevations and side slopes of the basins away from the streams would unduly bias the average parameter value for each stream. For example, radiation load directly on a stream would have a more significant influence on stream temperature than the average load on the stream basin. An example of simulated solar radiation extracted for the stream is shown in *figure 3* for August 26, for all streams considered in the stream temperature analysis. The variability of solar radiation load on streams in basins with steep topography is apparent; whereas, radiation load on streams in basins with flatter topography have less topographic shading and, thus, higher and less variable radiation loads. All distributed parameters (net radiation, relative humidity, and maximum and minimum daily air temperature) were extracted for the streams from the daily grids developed for all parameters to produce a daily time series for Jan. 1, 1999, to Dec. 31, 2008, of mean values for each of 24 stream basins (*fig. 3*).

Development of Regression Equations

An equation for stream temperature (StrmT) was developed for each tributary by using available intermittent maximum and minimum daily stream temperature data for 24 tributaries with measured stream temperature data, following the form:

$$StrmT = a + b(Rn) + c(VDD) + d(Tmean) + e(DA) + f(DA2) \qquad (4)$$

where

   Rn is basin-averaged daily net radiation,

   VDD is vapor density deficit,

   $Tmean$ is mean air temperature calculated as the mean of the maximum and minimum daily air temperature, and

   DA is a day of year function that accounts for the lag in earth temperature behind the maximum solar angle during the year.

For example, net radiation is at a maximum in late June when stream temperatures are still rising. This seasonality was accounted for by using a sine response to day of the year (from 0 to 360 degrees), then offset so the maximum (1) occurred in late summer, and the minimum (-1) occurred in mid-winter. This was calculated by using the equation:

Table 3. Locations of thermochrons used for stream temperature measurements in the Klamath River Basin.

[KBRT, Klamath Basin Rangeland Trust; PacifiCorp, electric power company; USFS, U.S. Forest Service; USGS, U.S. Geological Survey]

Stream identi-fication	Measurement location	Data source	Latitude, decimal degrees	Longitude, decimal degrees
		Lower Klamath Basin		
1	Trinity River near Weitchpec	Kris Web, 1998	41.1763	−123.6926
2	South Fork Trinity River	Kris Web, 1998	40.8786	−123.6011
3	North Fork Trinity River	Kris Web, 1998	40.8907	−123.5961
4	Salmon River at mouth	Yurok tribe	41.3751	−123.4848
5	Scott River at Fort Jones	Yurok tribe	41.7762	−123.0320
6	Shasta River at mouth	Yurok tribe	41.8197	−122.6057
		Upper Klamath Basin		
7	Jenny Creek near mouth	PacifiCorp	42.5847	−121.8486
8	Fall Creek near mouth	PacifiCorp	42.5514	−121.6186
9	Shovel Creek near mouth	PacifiCorp	42.4658	−121.5153
10	Spencer Creek near mouth	PacifiCorp/USFS	42.4600	−121.2700
11	Cherry Creek	USFS	42.4478	−121.2375
12	Sevenmile Creek near Dry Creek	USFS	42.4853	−121.0944
13	Crooked Creek at Root Ranch	KBRT	42.4969	−121.0050
14	Sprague River near Chiloquin	USGS	43.5489	−121.1147
15	Trout Creek below confluence	Klamath tribe	42.4319	−121.0161
16	Sprague River at Tinkers	Klamath tribe	42.3736	−120.8442
17	Sycan River at Elde Flat	Klamath tribe	42.3906	−120.9125
18	Sycan River at Drews Road	Klamath tribe	42.4850	−121.2772
19	Sprague River near Beatty	USGS/Klamath tribe	42.6153	−121.3461
20	Fivemile Creek	Klamath tribe	42.4881	−121.6178
21	North Fork Sprague River at Ivory Pine	Klamath tribe	42.6208	−121.9675
22	North Fork Sprague River at Elbow	Klamath tribe	42.6111	−121.9386
23	South Fork Sprague River at Campbell Road	Klamath tribe	42.7253	−122.0881
24	South Fork Sprague River at Blaisdell	Klamath tribe	42.5983	−122.0944

$$DA = sine\,(((DOY)\,/\,365)*360 + t) \qquad (5)$$

where

  DOY represents the day of year, and

   t is a fit parameter that offsets the time of mean
      stream temperatures and is fit at the same
      time as the other parameters.

Fits are done to minimize the root mean square error between observed and estimated stream temperature.

Correction and Application of Future Air Temperature Projections

The future temperature projections were originally bias-corrected following a two-step procedure described by Wood and others (2002) and Maurer (2007). The observed data that were used for the bias correction are daily gridded data that are available at 12-km spatial resolution from 1950–99 (Maurer and others, 2002). During development of the future-temperature files, it was noted that average temperatures for the historical baseline period 1950–99 were generally lower than the local weather-station data used to calibrate the fish production water-quality model. This station was located in Montague, CA, in the Shasta River Basin, had a sparse record, and was not used in the development of temperature grids for

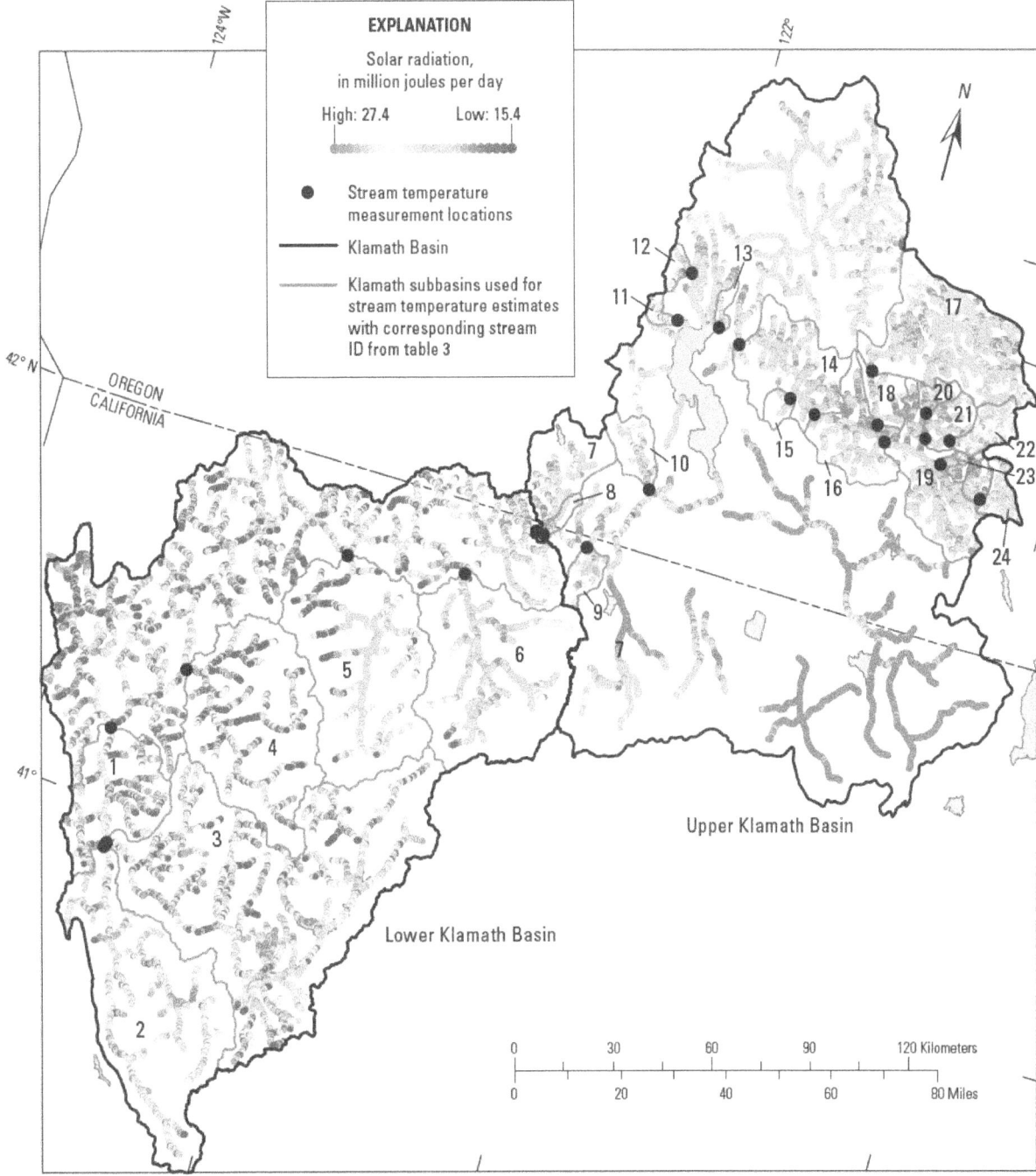

Figure 3. Spatially distributed solar radiation for August 26 for stream channels, all stream temperature measurement locations, and associated upstream basins used for stream temperature estimates with corresponding stream ID from *table 3*.

the stream-temperature modeling. The historical period of the temperature projections was compared to another historical gridded climate dataset, PRISM, which is available at a 4-km spatial resolution for monthly values (Parameter-Elevation Regressions on Independent Slopes Model; Daly and others, 1994). The results of this comparison indicated that the mean air temperature for the 50-year (yr) period for the PRISM dataset was 0.86°C higher for the Upper Klamath Basin, and 1.87°C higher for the Lower Klamath Basin, than the temperature projections provided by Reclamation (*table 4*) and more closely matched the locally measured air temperature. As a result, the five climate projections were corrected by this magnitude for the 2000–99 record.

For application of the basin-averaged future air-temperature projections, a method was developed to spatially distribute the daily air temperature throughout the basins. An adiabatic lapse rate (6.9°C per 1,000 meters) from the mean elevation for both the Upper and Lower Klamath Basins was calculated for every grid cell on the basis of a 270-meter (m) digital elevation model, and applied to each climate projection for each region. The resulting air temperature for each day was extracted for all stream cells upstream of measured stream temperature locations. These values were then used with the calibrated regression equations to estimate future stream temperatures.

As future projections only provided daily mean air temperature, estimates of future net radiation and vapor density deficit were also required. Net radiation was simulated by using future projections of air temperature for the calculation of long-wave radiation following equations in Flint and Flint (2008). Cloudy sky conditions were not incorporated into the estimate as a temperature range of maximum and minimum air temperatures were not available for future projections. To estimate vapor density deficit on the basis of mean air temperature, a polynomial equation was fit to the relation between mean air temperature and vapor density deficit for all streams. The equation was used with future air temperature projections extracted for each stream to calculate future vapor density deficit.

Results and Discussion

Regression Analysis

Regression equations were developed for each tributary stream, and the error between observed and estimated stream temperature was minimized on the basis of root mean square error. The equations to calculate vapor density deficit for each stream, all regression equation coefficients and statistics, and number of stream temperature measurements in the period of record are included in *table 5*. Coefficient of determination for the estimate of vapor density deficit on the basis of mean air temperature was generally high, with r^2 values between 0.7 and 0.9. The standard error of the y-estimate (SEE) for the estimation of stream temperature for the 24 streams ranged from 0.36°C to 1.64°C, with an average error of 1.12°C for all streams. Generally, the smaller basins have a lower SEE, often under a degree C (*table 5*). In addition, streams with longer measurement records appear to have a higher SEE (*table 5*). The quality of temperature record, whether continuous or spotty, or with occasional large spikes or dives in temperature, can also lead to higher errors in the estimate of stream temperature. Whether or not the measurement data provided bounds to the seasonal stream-temperature range was also a factor in the resulting error. The range of parameter coefficients in *table 5* is due to the lack of a physical basis for the purely empirical derivations. Regressions were developed specifically for each tributary on the basis of the measured data. The regressions were then applied to gaps in the data record at those sites to complete the 10-yr calibration period, and were then applied to the 100-yr future period. Stream temperature estimates for Spencer Creek, extrapolated to the calibration period, 1999–2008, on the basis of a regression developed from intermittent data, are shown in *figure 4*. Included in the figure are the measured versus predicted stream temperatures and regression equation, as well as the polynomial fit of vapor density deficit and mean daily air temperature.

Table 4. Comparison of air temperature projections and PRISM air temperature for 1950–99 to establish correction factors for future temperature projections for the Upper and Lower Klamath Basins.

	Projections		PRISM		Correction	
	Upper basin	Lower basin	Upper basin	Lower basin	Upper basin	Lower basin
Mean	5.42	8.35	6.28	10.22	0.86	1.87
Standard deviation	6.74	6.4	6.64	6.39		

Table 5. Regression equations and errors for stream temperature in streams of the Klamath River basin.

[DA, day of year function; r², coefficient of determination; RMSE, root mean square error; Rn, basin-averaged daily net radiation in million joules per day; SEE, standard error of estimate; Tmean, mean daily temperature; VDD, vapor density deficit in millibars; km², square kilometer]

Stream identification	Basin name	Basin area (km²)	Vapor density deficit equation	r²	Equation coefficients						RMSE	SEE	Days of record
					a	b (Rn)	c (VDD)	d (Tmean)	e (DA)	f (DA2)			
			Lower Klamath Basin										
1	Trinity River	863	$y = 0.0229x^2 - 0.0478x + 0.6952$	0.82	3.61	0.73	−0.02	0.21	3.54	−2.82	1.55	1.49	1,021
2	South Fork Trinity River	2,353	$y = 0.0193x^2 - 0.0407x + 0.3587$	0.85	1.08	0.69	−0.11	0.41	4.10	−2.92	1.24	1.21	1,099
3	North Fork Trinity River	4,455	$y = 0.0216x^2 - 0.0947x + 0.722$	0.86	−0.06	0.65	−0.08	0.40	4.19	−3.95	1.31	1.27	984
4	Salmon River	1,948	$y = 0.0145x^2 + 0.0666x + 0.6225$	0.78	−0.84	0.74	0.40	0.13	2.51	−1.29	1.44	1.58	1,275
5	Scott River	2,108	$y = 0.019x^2 - 0.0123x + 0.6729$	0.89	−2.08	0.62	−0.17	0.55	4.12	−1.44	1.49	1.54	285
6	Shasta River	2,037	$y = 0.0179x^2 - 0.0097x + 0.5846$	0.9	2.14	0.36	0.06	0.54	0.36	−1.40	0.83	0.98	502
			Upper Klamath Basin										
7	Jenny Creek	488	$y = 0.0184x^2 + 0.0227x + 1.1852$	0.86	0.35	0.43	−0.08	0.50	1.30	−0.95	0.82	0.82	860
8	Fall Creek	57	$y = 0.0177x^2 + 0.0434x + 1.0835$	0.87	6.68	0.08	0.03	0.21	0.05	−0.39	0.37	0.36	329
9	Shovel Creek	134	$y = 0.0185x^2 + 0.0035x + 1.0848$	0.86	2.92	0.25	−0.09	0.35	1.51	−1.02	0.54	0.53	853
10	Spencer Creek	223	$y = 0.0175x^2 + 0.014x + 0.9925$	0.85	−3.18	0.59	0.03	0.44	1.29	−1.47	1.17	1.16	1,198
11	Cherry Creek	51	$y = 0.0131x^2 + 0.0911x + 0.5249$	0.83	−0.49	0.39	−0.06	0.17	2.67	−2.52	1.18	0.90	1,209
12	Sevenmile Creek	68	$y = 0.0129x^2 + 0.0815x + 0.6218$	0.83	2.53	0.11	−0.01	0.13	0.51	−0.89	0.52	0.50	1,629
13	Crooked Creek	55	$y = 0.0125x^2 + 0.088x + 0.6455$	0.82	5.24	0.16	0.07	0.19	−0.16	−0.38	0.48	0.47	766
14	Sprague River near Chiloquin	750	$y = 0.0112x^2 + 0.1011x + 0.4998$	0.81	−2.26	1.07	−0.07	0.15	3.38	−2.22	1.47	1.64	1,635
15	Trout Creek	76	$y = 0.0137x^2 + 0.047x + 0.4507$	0.8	−0.67	0.67	−0.07	0.01	−1.09	−1.75	1.30	1.26	1,678
16	Sprague River at Tinkers	518	$y = 0.0126x^2 + 0.0507x + 0.4871$	0.76	−0.95	0.67	0.26	0.34	2.37	−1.14	1.34	1.32	1,275
17	Sycan River at Elde Flat	1,246	$y = 0.0131x^2 + 0.0585x + 0.6348$	0.78	−3.29	0.98	0.14	0.14	3.16	−2.17	1.48	1.54	1,580
18	Sycan River at Drews Road	186	$y = 0.0114x^2 + 0.076x + 0.582$	0.76	−0.14	0.63	0.31	0.24	2.22	−1.29	1.37	1.44	1,314
19	Sprague River near Beatty	635	$y = 0.0142x^2 + 0.006x + 0.6872$	0.7	1.97	0.60	0.09	0.21	2.83	−2.01	1.18	1.22	1,832
20	Fivemile Creek	83	$y = 0.012x^2 + 0.0535x + 0.7016$	0.77	8.11	0.32	0.03	0.05	1.74	−1.02	1.38	1.26	1,868
21	North Fork Sprague River at Ivory Pine	181	$y = 0.0117x^2 + 0.0472x + 0.6305$	0.7	−1.67	0.72	0.17	0.29	3.78	−4.26	1.26	1.24	1,355
22	North Fork Sprague River at Elbow	187	$y = 0.0139x^2 + 0.0524x + 0.6265$	0.74	1.72	0.29	0.00	0.25	0.97	−1.56	0.61	0.67	1,592
23	South Fork Sprague River at Campbell Road	63	$y = 0.0142x^2 + 0.021x + 0.6348$	0.72	−2.65	0.84	0.07	0.30	3.91	−3.50	1.33	1.42	1,622
24	South Fork Sprague River at Blaisdell	136	$y = 0.0134x^2 + 0.0233x + 0.6264$	0.73	−1.57	0.58	0.07	0.28	2.38	−3.07	0.98	1.09	1,493
											1.12		1,218.92

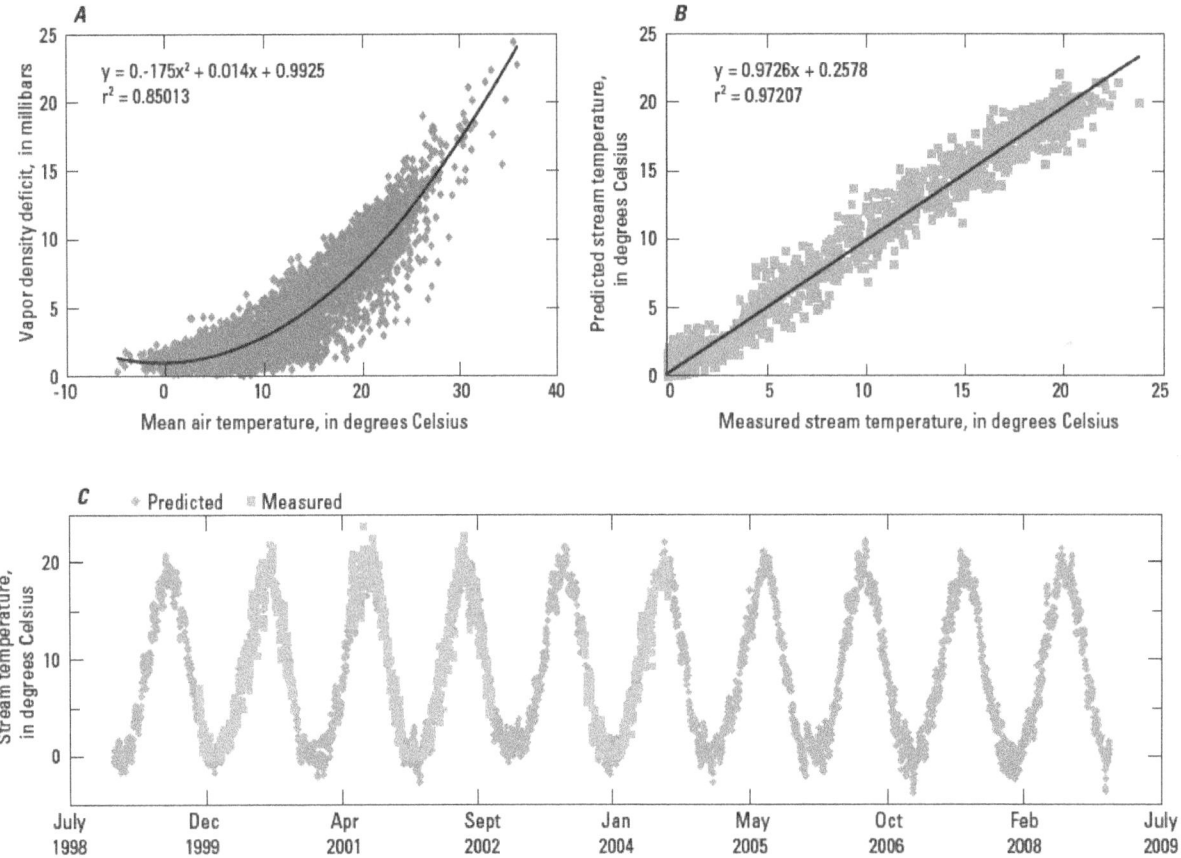

Figure 4.   Regression results for Spencer Creek illustrating (*A*) relation of vapor density deficit to mean air temperature, (*B*) relation of predicted and measured stream temperature, and (*C*) time series of predicted and measured stream temperature for Jan. 1, 1999, through Dec. 31, 2008.

Projected Future Stream Temperatures

Regression equations were used with future air-temperature projections to calculate both current stream temperatures for the calibration period, 1999–2008, and future projections of stream temperature for 2000–99 for all tributary streams, using all five air temperature projections. To illustrate the variation among just two of the projections for two tributary basins, the measured stream temperature, air temperature, and predicted stream temperature are accompanied by the projected air temperature and calculated stream temperature for Run 6 and Run 11 (*fig. 5*; *table 1*). For the Scott River, it can be seen that the projected air temperatures are slightly lower than the measured air temperatures, translating into lower stream temperatures for the projections. The measured and projected air temperatures for the Salmon River show more variability in the projections, rising above and falling

below the measured air temperature, which also translates into greater variability for the projected stream temperature, especially for Run 6.

The variation among projections for mean and maximum daily stream temperature, as well as the change between current and future air temperature conditions, is shown in *figure 6* and *table 6* for all streams. There are notable differences among the projections and among streams. For example, Run 45 is the lowest for the Shasta River stream temperatures (*fig. 6*; *table 6*), whereas Run 37 is the lowest for the North Fork and South Fork Trinity Rivers (*fig. 6*; *table 6*). Some streams show very little difference among projections, such as Fall, Shovel (*fig. 6*; *table 6*), and Trout Creeks (*fig. 6*; *table 6*). The change over the century is also variable, from over a 2°C change between baseline (1950–99) and the end of the century (Shasta River, *fig. 6*; *table 7*) to no change at all (Trout Creek, *fig. 6*; *table 7*).

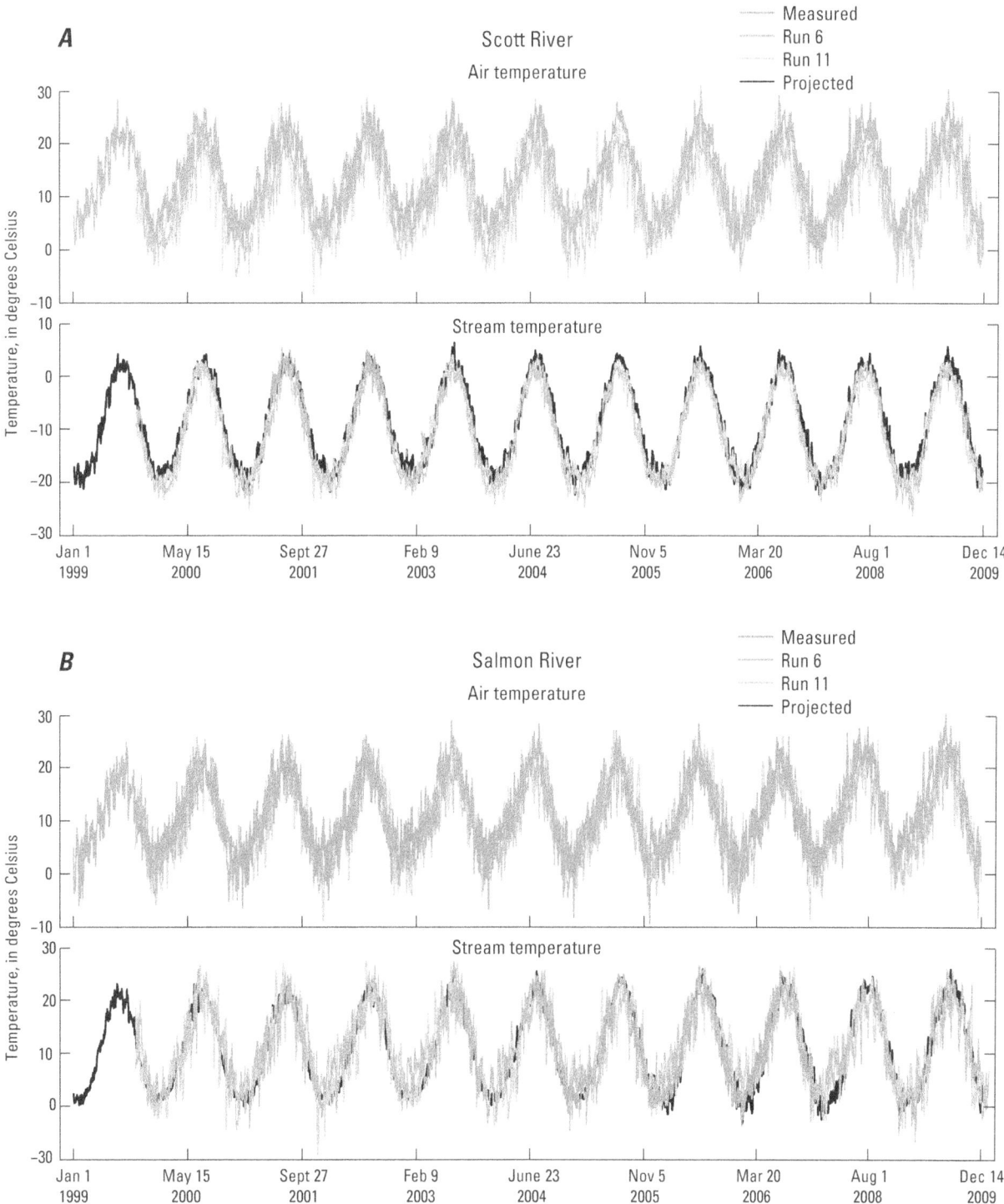

Figure 5. Measured and projected air temperature and stream temperature for 1999–2008 for the (*A*) Scott River and the (*B*) Salmon River, California.

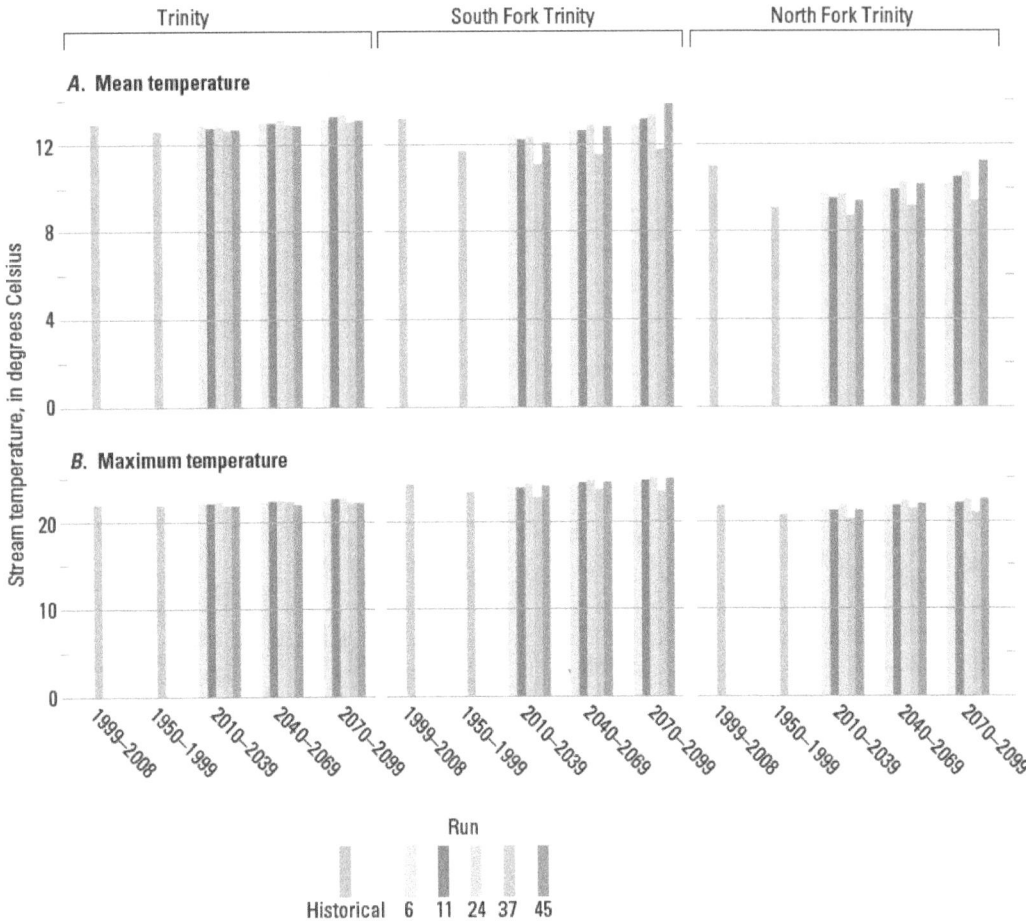

Figure 6. Histograms depicting the (*A*) mean daily stream temperature and (*B*) mean maximum daily stream temperature for 1950–99; the calibration period, 1999–2008; and future projections for 2011–40, 2041–70, and 2071–2100 for all streams in the Klamath River Basin.

Included in the figures and table are both the daily average mean (*fig. 6A*) and maximum stream temperature (*fig. 6B*) for 1999–2008, and the daily average (*fig. 6A*) and maximum baseline temperature (*fig. 6B*) for 1950–99. The baseline temperature was used to perform the bias correction, and, thus, the projections should exhibit change with the baseline as the starting point. This is apparent in the figures as the projected warming proceeds upward from the baseline through the end of the 21st century. However, there are numerous streams that exhibited a large change in temperature between the baseline period and the calibration period (1999–2008), such that most of the streams had higher temperatures and several of them had maximum-daily temperature increases of more than 2°C between the baseline and calibration periods (*table 7*). Of particular note are the Shasta River, which increased in mean daily stream temperature by 3.4°C, and the Salmon River, which increased in maximum daily stream temperature by 4.2°C (*table 7*). There are numerous streams that show little measured change over the last 58 years, particularly Crooked Creek, the Sprague River near Chiloquin, and Trout Creek (*table 7*), all of which are located relatively close together (*fig. 3*) and potentially moderated by large groundwater

inflows. Jenny and Fall Creeks, located farther west and also close together (*fig. 3*), exhibited small changes over time as well (*table 7*). There is no systematic moderation of the degree of change in stream temperature on the basis of dominance by groundwater flows in the Upper Basin, however, because several streams exhibited large changes, such as SF Sprague River at Blaisdell and Sprague River at Tinkers (*table 7*). The mean changes are lower for the Upper Basin, however, than for the Lower Basin. The large rivers in the Lower Basin show large increases in temperature between the baseline and calibration period, with the exception of the Trinity River, which had smaller changes in stream temperature than the Salmon, Scott, and Shasta Rivers. The Trinity River Basin experiences greater coastal temperature moderating effects, and has fewer land uses that enhance stream warming than the other larger rivers. For these three rivers, the largest change is for the Shasta and the smallest is for the Salmon, perhaps indicating coastal moderating of temperatures or different land-use practices. The mean change in stream temperature between the historical baseline period and the last 10 years is 1.2°C, and the mean change stream temperature between the baseline period and the end of the century is 1°C.

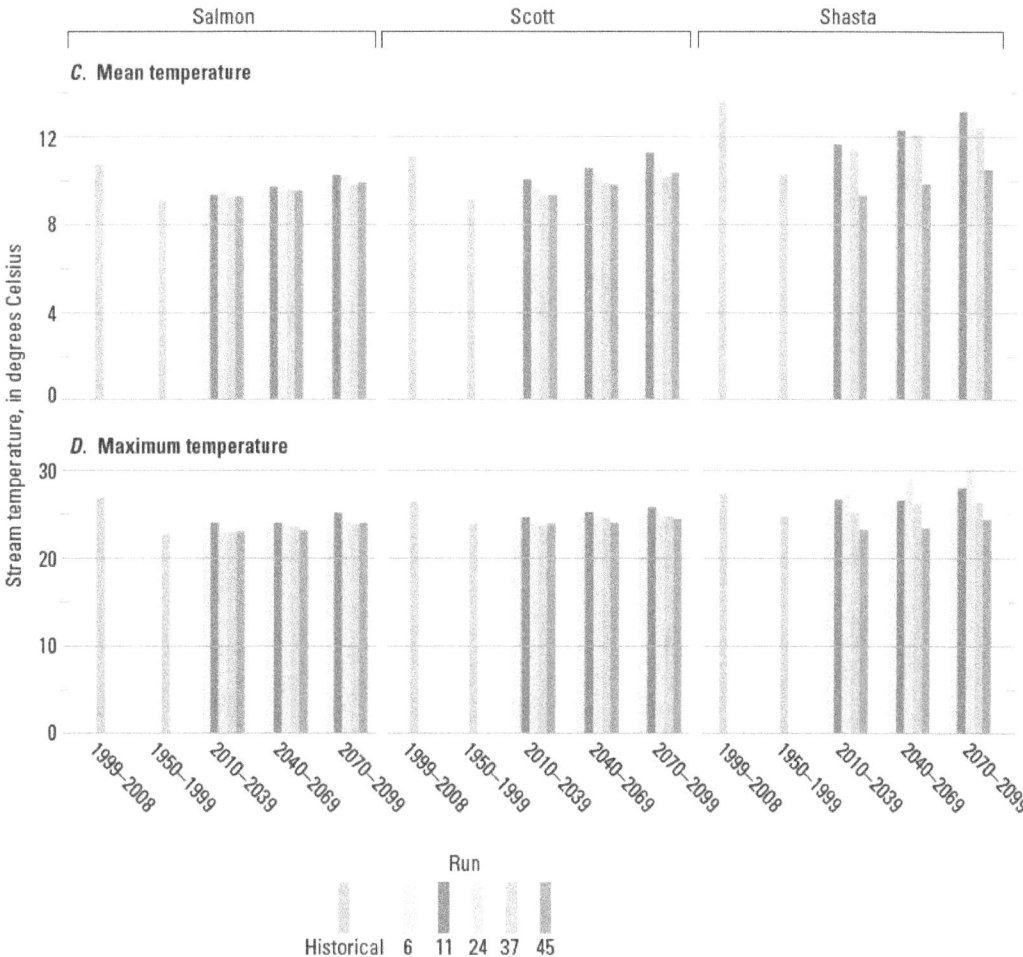

Figure 6.—Continued

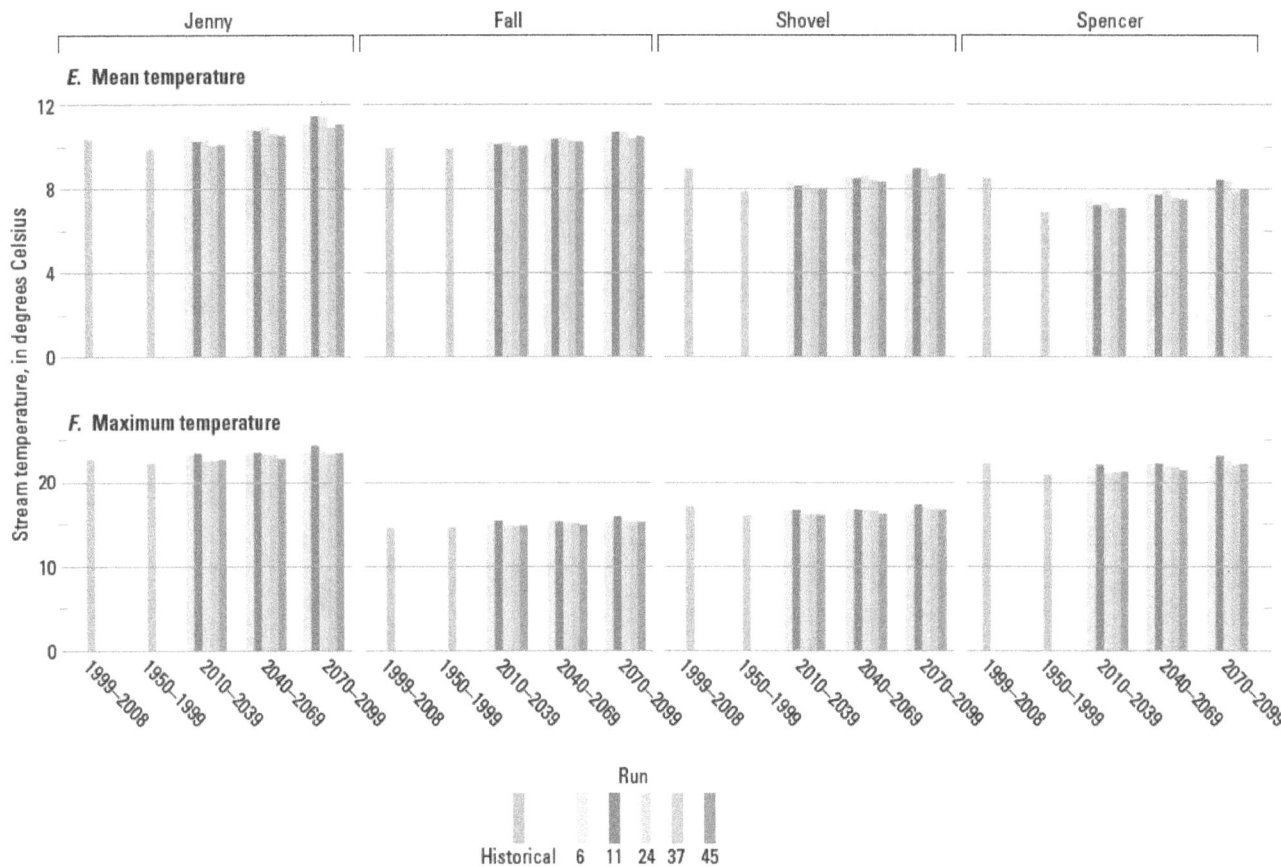

Figure 6.—Continued

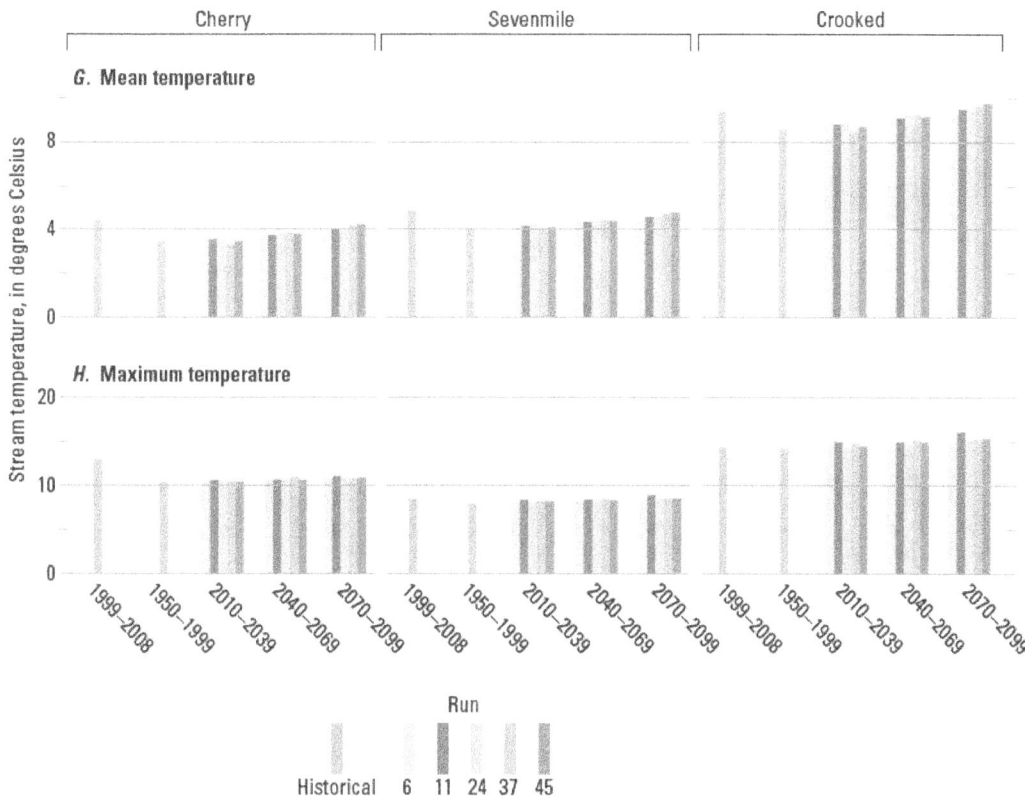

Figure 6.—Continued

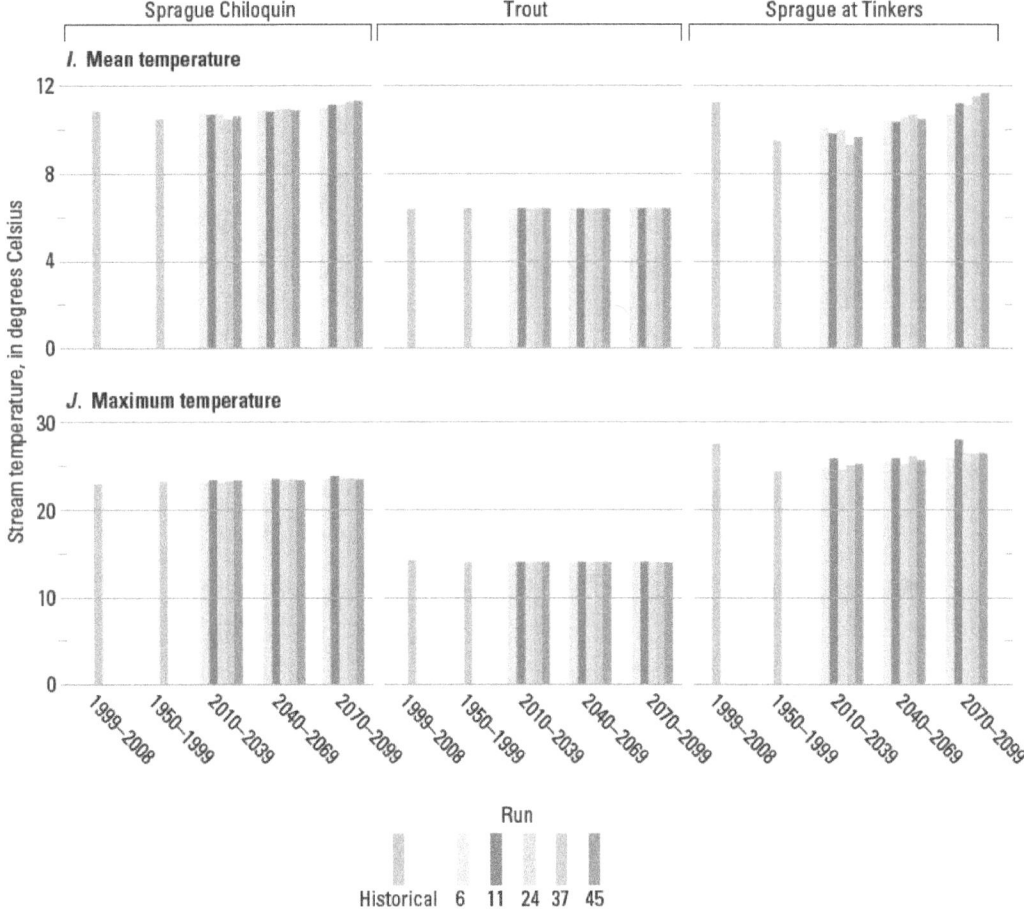

Figure 6.—Continued

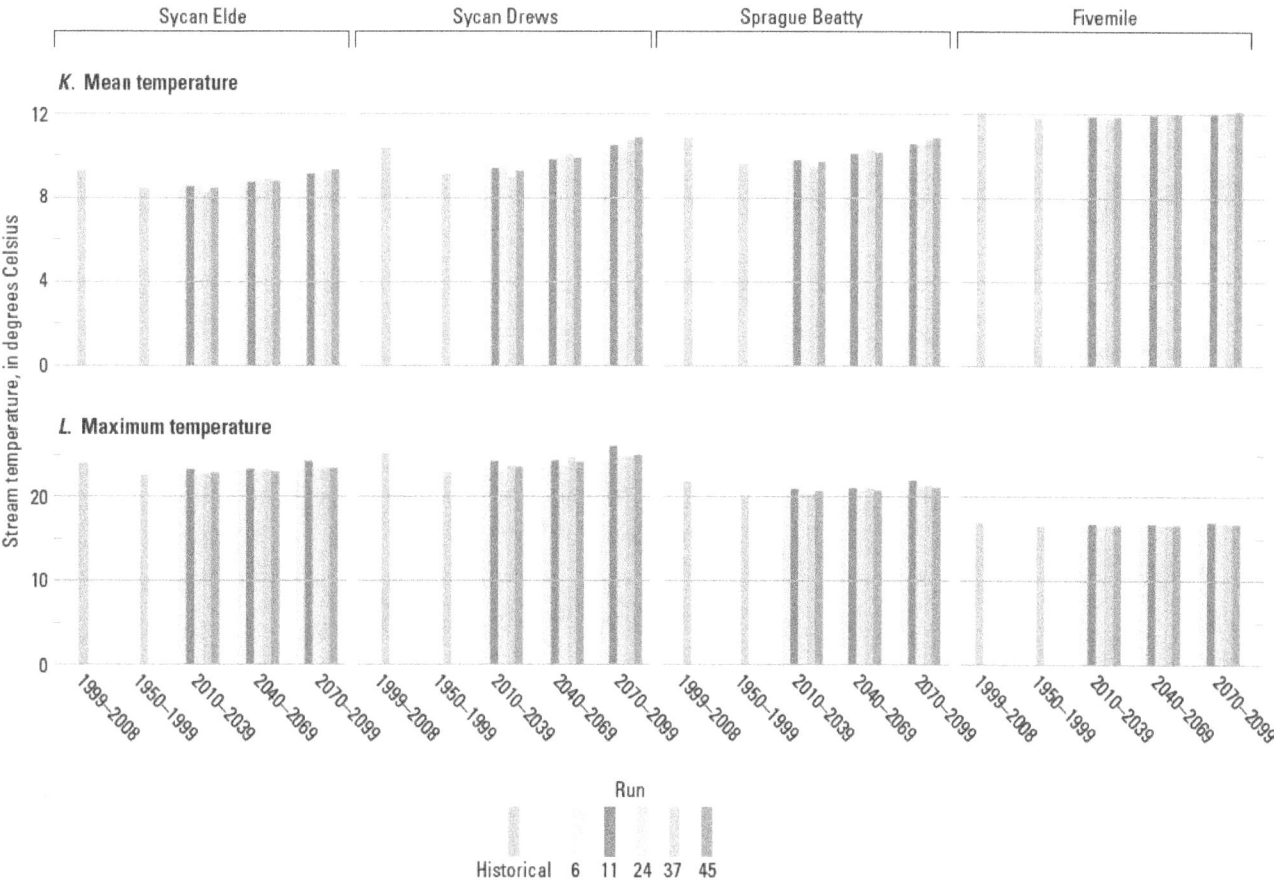

Figure 6.—Continued

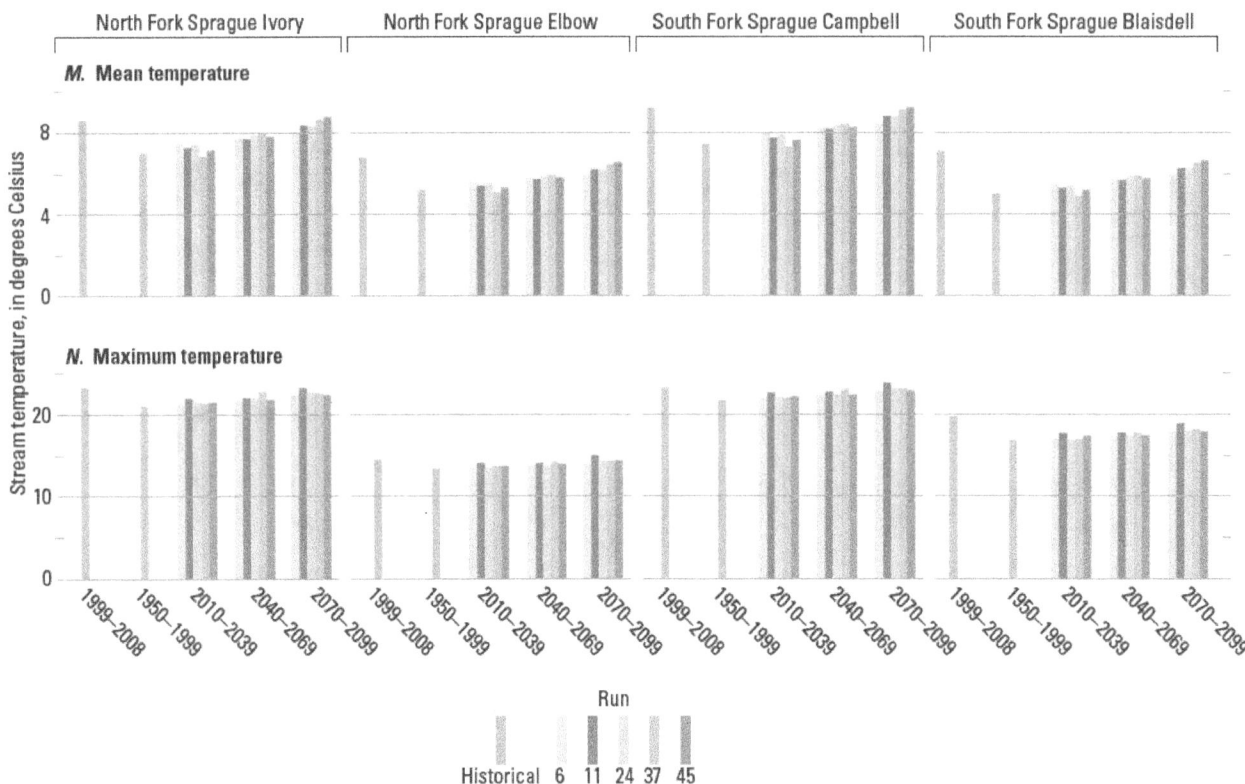

Figure 6.—Continued

Table 6. Mean and maximum daily stream temperature, in degrees Celsius, for two historical periods and three 30-year future periods for five future projections in and around the Klamath River Basin.

Stream	Time period	Mean daily stream temperature						Maximum daily stream temperature						Change in historical temperature (1999–2008)–(1950–2000)		Change in future temperature (2070–2099)–(1950–2000)	
		Run 6	Run 11	Run 24	Run 37	Run 45	Historical	Run 6	Run 11	Run 24	Run 37	Run 45	Historical	Mean	Maximum	Mean	Maximum
Trinity River	1999–2008						12.9						22.5	0.3	0.1	0.6	0.6
	1950–1999						12.6						22.4				
	2010–2039	12.9	12.7	12.8	12.7	12.7		22.7	22.6	22.8	22.3	22.4					
	2040–2069	13.0	13.0	13.1	12.9	12.9		22.8	22.9	23.0	22.8	22.4					
	2070–2099	13.1	13.3	13.4	13.0	13.1		23.0	23.2	23.3	22.7	22.7					
South Fork Trinity River	1999–2008						13.2						24.8	1.5	0.9	1.3	1.2
	1950–1999						11.7						23.9				
	2010–2039	12.4	12.2	12.4	11.1	12.1		24.6	24.4	24.9	23.3	24.6					
	2040–2069	12.6	12.6	12.9	11.5	12.8		24.8	25.0	25.3	24.2	25.0					
	2070–2099	12.9	13.1	13.3	11.8	13.8		25.0	25.3	25.6	24.0	25.5					
North Fork Trinity River	1999–2008						11.0						22.3	1.9	1.1	1.3	1.3
	1950–1999						9.1						21.2				
	2010–2039	9.7	9.5	9.7	8.7	9.4		21.9	21.7	22.4	20.8	21.8					
	2040–2069	10.0	9.9	10.3	9.2	10.2		22.2	22.3	22.9	22.0	22.5					
	2070–2099	10.2	10.5	10.7	9.4	11.2		22.4	22.6	23.0	21.5	23.0					
Salmon River	1999–2008						10.8						27.0	1.7	4.2	1.0	1.5
	1950–1999						9.1						22.8				
	2010–2039	9.6	9.4	9.5	9.3	9.3		23.7	24.1	23.0	23.0	23.1					
	2040–2069	9.8	9.7	9.9	9.6	9.6		24.1	24.1	23.9	23.7	23.2					
	2070–2099	10.0	10.3	10.2	9.8	9.9		24.2	25.2	24.4	23.9	24.1					
Scott River	1999–2008						11.1						26.5	1.9	2.5	1.4	1.1
	1950–1999						9.2						24.0				
	2010–2039	9.8	10.1	9.7	9.3	9.4		24.4	24.7	24.2	23.8	24.0					
	2040–2069	10.1	10.6	10.3	9.9	9.8		24.7	25.3	24.9	24.7	24.1					
	2070–2099	10.4	11.3	10.8	10.2	10.4		25.0	25.8	25.3	24.9	24.5					
Shasta River	1999–2008						13.7						27.5	3.4	2.6	2.1	2.3
	1950–1999						10.3						24.9				
	2010–2039	12.0	11.7	10.8	11.5	9.4		26.2	26.7	27.4	25.3	23.3					
	2040–2069	12.4	12.3	12.1	12.1	9.9		26.7	26.7	29.3	26.3	23.5					
	2070–2099	12.7	13.2	13.1	12.5	10.5		26.9	28.1	30.4	26.5	24.5					

Table 6. Mean and maximum daily stream temperature, in degrees Celsius, for two historical periods and three 30-year future periods for five future projections.—Continued

Stream	Time period	Mean daily stream temperature						Maximum daily stream temperature						Change in historical temperature (1999–2008)–(1950–2000)		Change in future temperature (2070–2099)–(1950–2000)	
		Run 6	Run 11	Run 24	Run 37	Run 45	Historical	Historical	Run 6	Run 11	Run 24	Run 37	Run 45	Mean	Maximum	Mean	Maximum
Jenny Creek	1999–2008						10.4	22.6						0.5	0.4	1.3	1.4
	1950–1999						9.9	22.2									
	2010–2039	10.5	10.2	10.4	10.1	10.1			23.1	23.3	22.4	22.5	22.6				
	2040–2069	10.8	10.8	11.0	10.6	10.5			23.4	23.4	23.2	23.1	22.7				
	2070–2099	11.1	11.5	11.4	10.9	11.1			23.5	24.2	23.6	23.3	23.4				
Fall Creek	1999–2008						9.9	14.5						0.0	-0.1	0.7	0.8
	1950–1999						9.9	14.6									
	2010–2039	10.2	10.1	10.2	10.0	10.0			15.1	15.3	14.8	14.7	14.8				
	2040–2069	10.4	10.4	10.5	10.3	10.2			15.3	15.3	15.2	15.1	14.9				
	2070–2099	10.5	10.7	10.7	10.4	10.5			15.4	15.9	15.4	15.2	15.3				
Shovel Creek	1999–2008						9.0	17.1						1.1	1.1	0.9	0.9
	1950–1999						7.9	16.0									
	2010–2039	8.3	8.2	8.3	8.0	8.1			16.5	16.6	16.2	16.1	16.1				
	2040–2069	8.5	8.5	8.7	8.4	8.4			16.7	16.6	16.7	16.5	16.2				
	2070–2099	8.7	9.0	9.0	8.6	8.7			16.8	17.3	16.9	16.7	16.6				
Spencer Creek	1999–2008						8.5	22.2						1.6	1.4	1.3	1.5
	1950–1999						6.9	20.8									
	2010–2039	7.5	7.3	7.4	7.1	7.1			21.8	22.0	21.0	21.1	21.2				
	2040–2069	7.8	7.7	7.9	7.6	7.5			22.1	22.1	21.8	21.8	21.3				
	2070–2099	8.1	8.4	8.4	7.9	8.0			22.2	23.0	22.4	22.0	22.1				
Cherry Creek	1999–2008						4.4	13.0						0.9	2.6	0.6	0.5
	1950–1999						3.5	10.4									
	2010–2039	3.7	3.6	3.6	3.4	3.5			10.5	10.7	10.6	10.5	10.5				
	2040–2069	3.8	3.8	3.9	3.9	3.8			10.7	10.7	10.8	11.0	10.7				
	2070–2099	3.9	4.0	4.1	4.2	4.3			10.8	11.1	11.0	10.9	10.9				
Sevenmile Creek	1999–2008						4.9	8.5						0.8	0.5	0.6	0.6
	1950–1999						4.1	8.0									
	2010–2039	4.3	4.2	4.3	4.0	4.1			8.2	8.4	8.1	8.3	8.3				
	2040–2069	4.4	4.4	4.4	4.5	4.4			8.3	8.4	8.3	8.5	8.4				
	2070–2099	4.5	4.6	4.6	4.8	4.8			8.4	8.9	8.6	8.6	8.6				

Table 6. Mean and maximum daily stream temperature, in degrees Celsius, for two historical periods and three 30-year future periods for five future projections.—Continued

Stream	Time period	Mean daily stream temperature						Maximum daily stream temperature						Change in historical temperature (1999–2008)–(1950–2000)		Change in future temperature (2070–2099)–(1950–2000)	
		Run 6	Run 11	Run 24	Run 37	Run 45	Historical	Run 6	Run 11	Run 24	Run 37	Run 45	Historical	Mean	Maximum	Mean	Maximum
Crooked Creek	1999–2008						9.4						14.4	0.8	0.1	0.9	1.1
	1950–1999						8.6						14.3				
	2010–2039	8.9	8.8	8.9	8.5	8.7		14.7	15.0	14.4	14.9	14.6					
	2040–2069	9.1	9.1	9.2	9.3	9.2		14.9	15.0	14.7	15.2	15.0					
	2070–2099	9.3	9.5	9.5	9.7	9.8		14.9	16.1	15.3	15.3	15.4					
Sprague River near Chiloquin	1999–2008						10.9						22.9	0.4	–0.3	0.7	0.4
	1950–1999						10.5						23.2				
	2010–2039	10.8	10.7	10.7	10.5	10.6		23.2	23.4	23.1	23.3	23.4					
	2040–2069	10.9	10.8	10.9	11.0	10.9		23.4	23.5	23.4	23.5	23.3					
	2070–2099	11.0	11.1	11.1	11.3	11.3		23.5	23.8	23.6	23.5	23.5					
Trout Creek	1999–2008						6.4						14.2	0.0	0.2	0.0	0.0
	1950–1999						6.4						14.0				
	2010–2039	6.4	6.4	6.4	6.4	6.4		14.0	14.0	14.0	14.0	14.0					
	2040–2069	6.4	6.4	6.4	6.4	6.4		14.0	14.0	14.0	14.0	14.0					
	2070–2099	6.4	6.4	6.4	6.4	6.4		14.0	14.0	14.0	14.0	13.9					
Sprague River at Tinkers	1999–2008						11.2						27.5	1.7	3.1	1.7	2.3
	1950–1999						9.5						24.4				
	2010–2039	10.1	9.8	10.0	9.3	9.7		24.8	25.8	24.6	25.0	25.2					
	2040–2069	10.4	10.4	10.6	10.7	10.5		25.5	25.9	25.3	26.2	25.6					
	2070–2099	10.7	11.2	11.1	11.5	11.7		26.0	28.0	26.5	26.4	26.5					
Sycan River at Elde Flat	1999–2008						9.3						23.9	0.9	1.3	0.8	1.0
	1950–1999						8.4						22.6				
	2010–2039	8.6	8.5	8.6	8.3	8.5		22.6	23.3	22.7	22.8	22.9					
	2040–2069	8.8	8.8	8.9	8.9	8.8		23.1	23.3	23.0	23.3	23.0					
	2070–2099	8.9	9.1	9.1	9.3	9.3		23.3	24.3	23.7	23.5	23.4					
Sycan River at Drews Road	1999–2008						10.4						25.2	1.3	2.2	1.4	2.0
	1950–1999						9.1						23.0				
	2010–2039	9.6	9.4	9.5	9.0	9.3		23.5	24.3	23.2	23.7	23.6					
	2040–2069	9.9	9.8	10.0	10.1	9.9		24.0	24.3	23.8	24.7	24.2					
	2070–2099	10.1	10.5	10.5	10.8	10.9		24.2	26.0	24.8	24.8	25.0					
Sprague River near Beatty	1999–2008						10.8						21.9	1.2	1.7	1.0	1.2
	1950–1999						9.6						20.2				
	2010–2039	9.9	9.8	9.9	9.5	9.7		20.4	21.0	20.4	20.3	20.7					
	2040–2069	10.1	10.1	10.2	10.3	10.2		20.6	21.0	20.7	21.1	20.8					
	2070–2099	10.3	10.6	10.6	10.8	10.9		21.1	22.0	21.2	21.4	21.1					

Table 6. Mean and maximum daily stream temperature, in degrees Celsius, for two historical periods and three 30-year future periods for five future projections.—Continued

Stream	Time period	Mean daily stream temperature						Maximum daily stream temperature						Change in historical temperature (1999–2008)–(1950–2000)		Change in future temperature (2070–2099)–(1950–2000)	
		Run 6	Run 11	Run 24	Run 37	Run 45	Historical	Run 6	Run 11	Run 24	Run 37	Run 45	Historical	Mean	Maximum	Mean	Maximum
Fivemile Creek	1999–2008						12.1						17.0	0.3	0.4	0.2	0.2
	1950–1999						11.8						16.6				
	2010–2039	11.9	11.9	11.9	11.8	11.8		16.6	16.8	16.6	16.6	16.7					
	2040–2069	11.9	11.9	12.0	12.0	11.9		16.7	16.8	16.7	16.7	16.7					
	2070–2099	12.0	12.0	12.0	12.1	12.1		16.8	17.0	16.8	16.8	16.8					
North Fork Sprague River at Ivory Pine	1999–2008						8.6						23.4	1.6	2.3	1.4	1.6
	1950–1999						7.0						21.1				
	2010–2039	7.5	7.2	7.4	6.8	7.1		21.3	22.0	21.6	21.4	21.5					
	2040–2069	7.7	7.7	7.9	7.9	7.8		21.8	22.1	22.0	22.8	21.8					
	2070–2099	7.9	8.3	8.3	8.6	8.7		22.4	23.3	22.7	22.6	22.4					
North Fork Sprague River at Elbow	1999–2008						6.8						14.6	1.6	1.2	1.0	1.1
	1950–1999						5.2						13.4				
	2010–2039	5.5	5.4	5.5	5.0	5.3		13.6	14.1	13.5	13.7	13.7					
	2040–2069	5.7	5.7	5.9	5.9	5.8		13.9	14.1	13.8	14.3	14.0					
	2070–2099	5.9	6.2	6.2	6.4	6.5		14.0	15.1	14.4	14.4	14.4					
South Fork Sprague River at Campbell Road	1999–2008						9.2						23.3	1.8	1.6	1.4	1.5
	1950–1999						7.4						21.7				
	2010–2039	7.9	7.7	7.9	7.3	7.6		22.0	22.7	22.2	22.0	22.2					
	2040–2069	8.2	8.1	8.3	8.4	8.2		22.3	22.8	22.5	23.1	22.5					
	2070–2099	8.4	8.8	8.8	9.1	9.2		22.9	23.9	23.2	23.2	23.0					
South Fork Sprague River at Blaisdell	1999–2008						7.1						19.8	2.1	2.9	1.3	1.3
	1950–1999						5.0						16.9				
	2010–2039	5.5	5.3	5.4	4.9	5.2		17.0	17.7	16.9	17.0	17.4					
	2040–2069	5.7	5.7	5.8	5.9	5.7		17.3	17.8	17.4	17.8	17.5					
	2070–2099	5.9	6.2	6.5	6.5	6.6		17.9	18.9	18.0	18.2	17.9					

Table 7.   Changes in daily mean and maximum stream temperatures, in degrees Celsius (°C), from the 1950–99 baseline period to the 1999–2008 calibration period and 2070–99 projected period in the Klamath River Basin.

Stream identi-fication	Stream	Change in historical temperature (1999–2008)–(1950–1999)		Change in future temperature (2070–2099)–(1950–2000)	
		Mean daily temperature	Maximum daily temperature	Mean daily temperature	Maximum daily temperature
	Lower Basin				
1	Trinity River	0.3	0.1	0.6	0.6
2	South Fork Trinity River	1.5	0.9	1.3	1.2
3	North Fork Trinity River	1.9	1.1	1.3	1.3
4	Salmon River	1.7	4.2	1.0	1.5
5	Scott River	1.9	2.5	1.4	1.1
6	Shasta River	3.4	2.6	2.1	2.3
	Mean	1.8	1.9	1.3	1.3
	Upper Basin				
7	Jenny Creek	0.5	0.4	1.3	1.4
8	Fall Creek	0.0	-0.1	0.7	0.8
9	Shovel Creek	1.1	1.1	0.9	0.9
10	Spencer Creek	1.6	1.4	1.3	1.5
11	Cherry Creek	0.9	2.6	0.6	0.5
12	Sevenmile Creek	0.8	0.5	0.6	0.6
13	Crooked Creek	0.8	0.1	0.9	1.1
14	Sprague River near Chiloquin	0.4	-0.3	0.7	0.4
15	Trout Creek	0.0	0.2	0.0	0.0
16	Sprague River at Tinkers	1.7	3.1	1.7	2.3
17	Sycan River at Elde Flat	0.9	1.3	0.8	1.0
18	Sycan River at Drews Rd	1.3	2.2	1.4	2.0
19	Sprague River near Beatty	1.2	1.7	1.0	1.2
20	Fivemile Creek	0.3	0.4	0.2	0.2
21	North Fork Sprague River at Ivory Pine	1.6	2.3	1.4	1.6
22	North Fork Sprague River at Elbow	1.6	1.2	1.0	1.1
23	South Fork Sprague River at Campbell Road	1.8	1.6	1.4	1.5
24	South Fork Sprague River at Blaisdell	2.1	2.9	1.3	1.3
	Mean	1.0	1.3	1.0	1.1
	Total maximum	3.4	4.2	2.1	2.3
	Total minimum	0.0	-0.3	0.0	0.0
	Total mean	1.2	1.4	1.0	1.2

Summary and Conclusions

In support of fish production modeling for an upcoming Secretarial Determination regarding the removal of four dams in the Klamath River Basin, stream temperature estimates were required for the development of models projecting stream temperature and water quality for the 21st century. To take advantage of downscaled future mean daily air temperature projections from five emissions scenarios that were available from the U.S. Bureau of Reclamation, the method of Flint and Flint (2008) was employed to estimate stream temperatures. The method requires the development of spatially distributed maps of net solar radiation, relative humidity, and air temperature from which daily basin averages were modeled for grid cells that intersect tributary streams. These data were used along with measured stream temperatures for 1999–2008 to develop and calibrate regression equations for 24 selected streams in the Klamath River Basin. Mean standard error of the y-estimate for all regression equations was 1.12°C. The projections of future air temperature were further bias corrected and downscaled by using adiabatic lapse rates to produce daily maps from which future air temperature times series were modeled for stream grid cells. These future air temperatures were then used with the regression equations to project five sets of future daily stream temperatures for each of the 24 streams for 2010–99.

Projected stream temperatures varied among scenarios by as much as 2°C for some streams and not at all for other streams, indicating spatial variation for projected stream temperature changes in the Klamath River Basin. The baseline period, 1950–99, to which the air temperature projections were corrected, established the starting point for the projected changes in air temperature. The average measured daily air temperature for the calibration period, 1999–2008, however, was found to be as much as 4.2°C higher than the baseline for some rivers, indicating warming conditions have already occurred in many areas of the Klamath River Basin, and that the stream temperature projections for the 21st century could be underestimating the actual change.

Uncertainties in the applied approach to estimating future stream temperatures include more than the uncertainties associated with the future air-temperature projections. The projected stream temperatures rely on calibrations that might not reflect conditions found in the future, or might not incorporate processes that could play a larger role in the future than they do now. For example, changes in the timing of springtime snowmelt could alter seasonal stream temperatures in a manner not reflected by the calibration period, or could influence the ratio of groundwater to the total flow, and, thus change, its influence on the resulting stream temperature. Additionally, potential changes in streamflow are not considered in this application. Although these uncertainties should be considered for application of future climate projections, the relative increases in air temperature and associated increases in stream temperature, represented by the range of scenarios in this study, are likely to be a conservative estimate of the variability and extremes of future stream temperatures in the Klamath River Basin.

References Cited

Barnett, T., Malone, R., Pennell, W., Stammer, D., Semtner, B., and Washington, W., 2004, The effects of climate change on water resources in the West: Introduction and overview: Climatic Change v. 62, p. 1–11.

Bartholow, J.M., 2005, Recent water temperature trends in the lower Klamath River, California: North American Journal of Fisheries Management v. 25, p. 152–162.

Bartholow, J.M, Heasely, J., Hanna, R.B., Sandelin, J., Flug, M., Campbell, S., Henriksen, J., and Douglas, A., 2005, Evaluating water management strategies with the Systems Impact Assessment Model: SIAM Version 4. Revised October 2005: U.S. Geological Survey Open File Report 03–82, 133 p.

Brekke, L., Maurer, E., Anderson, J., Dettinger, M., Townsley, E., Harrison, A., and Pruitt, T., 2009, Assessing reservoir operations risk under climate change: Water Resources Research, v. 45, no. W04411, 16 p.

Campbell, G.S., 1979, An Introduction to Environmental Biophysics: New York, Springer-Verlag Inc., 159 p.

Campbell, G.S., and Norman, J.M., 1998, An Introduction to Environmental Biophysics: Springer-Verlag Inc., 286 p.

Cayan, D.R., Maurer, E.P., Dettinger, M.D., Tyree, M., and Hayhoe, K., 2008, Climate change scenarios for the California region: Climatic Change 87(Suppl 1), p. S21–S42.

Daly, C., Neilson, R.P., Phillips, D.L., 1994, A statistical-topographic model for mapping climatological precipitation over mountainous terrain: Journal of Applied Meteorology, v. 33, p. 140–158.

Dunne, T., Ruggerone, G., Goodman, D., Rose, K., Kimmerer, W., and Ebersole, J., 2011, Scientific assessment of two dam removal alternatives on Coho salmon and steelhead: Portland, Oregon, Final Report to the Klamath River Expert Panel, April 25, 216 p.

Flint, L.E., and Flint, A.L., 2008, A basin-scale approach to estimating stream temperature of tributaries at the Lower Klamath River, CA, California: Journal of Environmental Quality, v. 37, p. 57–68.

International Panel on Climate Change, 2000, Emissions scenarios *in* Nakicenovic, Nebojsa and Swart, Rob, eds., Special report of the Intergovernmental Panel on Climate Change: Cambridge, United Kingdom and New York, Cambridge University Press, 570 p.

Maurer, E.P., 2007, Uncertainty in hydrologic impacts of climate change in the Sierra Nevada, California, under two emissions scenarios: Climatic Change, v. 82, p. 309–325.

Maurer, E., and Duffy, P., 2005, Uncertainty in projections of streamflow changes due to climate change in California: Geophysical Research Letters, v. 32, no. L03704, 5 p.

Maurer, E.P., Wood, A.W., Adam, J.C., Lettenmaier, D.P., and Nijssen, B., 2002, A long-term hydrologically-based data set of land surface fluxes and states for the conterminous United States: Journal Climate, v. 15, no. 22, p. 3237–3251.

Meehl, G.A., Covey, C., Delsworth, T., Latif, M., McAvaney, B., Mitchell, J.F.B., Stouffer, R.J., and Taylor, K.E., 2007, The WCRIP CMIP3 multimodel dataset—A new era in climate change research: Bulletin of the American Meteorological Society, p. 1383–1394 (DOI:10.1175/BAMS-88-9-1383).

Mohseni, O., and Stefan, H.G., 1999, Stream temperature/air temperature relationship: a physical interpretation: Journal of Hydrology, v. 218, p. 128–141.

Nalder, I.A., and Wein, R.W.,1998, Spatial interpolation of climatic normals: Test of a new method in the Canadian boreal forest: Agricultural and Forest Meteorology, v. 92, p. 211–225.

National Climatic Data Center, 2009, Summary of the day observations, 3200-series data West 1: California, Nevada, Utah, Arizona, New Mexico, Colorado, and Wyoming: Boulder, Colorado, EarthInfo Inc.

National Marine Fisheries Service, 2006, National Marine Fisheries Service 10(j) recommendations for the Klamath River hydropower project FERC re-licensing: 56 p. (*http://swr.nmfs.noaa.gov/klamath/klam_habitat.htm*)

Payne, J., Wood, A., Hamlet, A., Palmer, R., and Lettenmaier, D., 2004, Mitigating the effects of climate change on the water resources of the Columbia River Basin: Climatic Change v. 62, p. 233–256.

Raff, David, 2009, Draft Klamath River Basin climate change projections and watershed model selection: U.S. Bureau of Reclamation white paper, October 24, 12 p.

Shuter, B.J., and Meisner, J.D., 1992, Tools for assessing the impact of climate change on freshwater fish populations: GeoJournal, v. 28, no. 1, p. 7–20.

U.S. Department of Interior, Bureau of Reclamation, San Joaquin River Restoration Program, 2009, Sensitivity of Future Central Valley Project and State Water Project Operations to Potential Cliamte Change and Associated Sea Level Rise: First Administrative Draft supplemental Hydrologic and Water Operations Analyses, Appendix, 110 p.

U.S. Department of Interior, Bureau of Reclamation, 2011, Hydrology, hydraulics and sediment transport studies for the Secretary's Determination on Klamath River dam removal and basin restoration: Denver, Colorado, U.S. Bureau of Reclamation, Mid-Pacific Region, Technical Service Center, Technical Report No. SRH-2011-02.

Vicuna, S., and Dracup, J., 2007, The evolution of climate change impact studies on hydrology and water resources in California: Climatic Change, v. 82, p. 327–350.

Wood, A.W., Leung, L.R., Sridhar, V., and Lettenmaier, D.P., 2004, Hydrologic implications of dynamical and statistical approaches to downscaling climate model outputs: Climatic Change, v. 15, no. 62, p. 189–216.